AF778192

# Urban Social Listening

Justin B. Hollander • Erin Graves • Henry Renski
Cara Foster-Karim • Andrew Wiley • Dibyendu Das

# Urban Social Listening

## Potential and Pitfalls of Using Microblogging Data in Studying Cities

palgrave
macmillan

Justin B. Hollander
Tufts University
Medford, Massachusetts, USA

Erin Graves
Urban Planner and Sociologist
Cambridge, USA

Henry Renski
University of Massachusetts Amherst
Amherst, Massachusetts, USA

Cara Foster-Karim
Tufts University
Medford, Massachusetts, USA

Andrew Wiley
Tufts University
Medford, Massachusetts, USA

Dibyendu Das
Tufts University
Medford, Massachusetts, USA

ISBN 978-1-137-59490-7 ISBN 978-1-137-59491-4 (eBook)
DOI 10.1057/978-1-137-59491-4

Library of Congress Control Number: 2016948422

© The Editor(s) (if applicable) and The Author(s) 2016
The author(s) has/have asserted their right(s) to be identified as the author(s) of this work in accordance with the Copyright, Designs and Patents Act 1988.
This work is subject to copyright. All rights are solely and exclusively licensed by the Publisher, whether the whole or part of the material is concerned, specifically the rights of translation, reprinting, reuse of illustrations, recitation, broadcasting, reproduction on microfilms or in any other physical way, and transmission or information storage and retrieval, electronic adaptation, computer software, or by similar or dissimilar methodology now known or hereafter developed.
The use of general descriptive names, registered names, trademarks, service marks, etc. in this publication does not imply, even in the absence of a specific statement, that such names are exempt from the relevant protective laws and regulations and therefore free for general use.
The publisher, the authors and the editors are safe to assume that the advice and information in this book are believed to be true and accurate at the date of publication. Neither the publisher nor the authors or the editors give a warranty, express or implied, with respect to the material contained herein or for any errors or omissions that may have been made.

Cover illustration: Modern building window © saulgranda/Getty

Printed on acid-free paper

This Palgrave Macmillan imprint is published by Springer Nature
The registered company is Macmillan Publishers Ltd. London

# Acknowledgments

We are very grateful to Tufts University's Office of the Vice Provost through the Tufts Collaborates! Grant Program for initial support of this research and to our colleagues on that book project, Profs. Tama Leventhal and Natalya Zaika. More advanced stages of the research from this book were supported by the Lincoln Institute of Land Policy (grant UJH080714); special thanks go to Armando Carbonell. We appreciate the generosity of IBM in providing us with a gratis one-year license of SPSS Modeler; special thanks also go to Dr. Jeffrey Nichols at IBM for his early help in downloading tweets.

Early versions of some of these chapters were presented at The Big Data and Urban Informatics Workshop, University of Illinois (August 2014), and at the Association of Collegiate Schools of Planning Annual Conference in Philadelphia (October 2014). We are grateful for the valuable advice and suggestions offered by moderators and others at those sessions. We also appreciate the early advice of Simcha Leventhal and Donovan Guin to get this project going.

We also want to acknowledge the efforts of our many colleagues at Tufts for helping to make this research happen, including Jackie Dejean, Leigh Curley, Susan Morrison, Ann Urosevich, Maria Nicolau, and Mike Flanary. Research assistance was provided by the following Tufts students: Haleigh Williams, Owen Searls, Evan Sullivan, Ryan Bailey, and Suveer Bahirwani.

Most of all, we want to express our deep appreciation for the love and support our families have given us that has made this book possible: thank you!

# Contents

# List of Figures

# List of Tables

CHAPTER 1

# Introduction

**Abstract** Every day, millions of people around the globe express themselves on social media, generating large, complex, and intriguing datasets. Academics, policy makers, pollsters, and firms are rushing to interpret this digital self-expression to answer a wide variety of questions. Our research is an attempt to analyze sentiment in microblogs to learn more about places and the people that occupy them. This book is an attempt to chronicle that work in service of demonstrating to planners and policy makers how and how not to use Big Data to understand sentiment, populations, and places.

**Keywords** Big Data • Sentiment analysis • Content analysis • Microblogs • Twitter

Given the extraordinary volume of data emerging from social media, urbanists have a unique opportunity to gain enhanced insights into the attitudes and opinions of people in cities. While technologies are ever changing and "disruption is the new norm," platforms like YouTube, Facebook, Instagram, Snapchat, and blogging demonstrate the breadth and durability of digital self-expression. Twitter, in particular, is among the most established and mainstream of microblogging venues. It is also inherently populist, with everyone from heads of church and state to heads of households, and individuals who make up those households, possessing Twitter accounts. It is

© The Editor(s) (if applicable) and The Author(s) 2016

J.B. Hollander et al., *Urban Social Listening*,
DOI 10.1057/978-1-137-59491-4_1

the self-expression of this great breadth of individuals that we find most compelling, and our efforts here are geared toward making evident the ways in which research using large social media data sets provides an additional investigative tool for planners and policy makers, even as we provide a candid discussion of the ways in which this method might be improved.

Thus, this book seeks to show how Twitter, as a prominent case of digital self-expression, can be analyzed and contribute to our understanding of place. Moreover, this book explores the utility of a new and potentially robust data source: the Twitter Application Program Interface (API) feed. We provide some context for the broader use of social media in urban studies and planning, explore the strengths and weaknesses of data gleaned from Twitter, and then demonstrate how such data can be used in research and practice. The three demonstration studies were completed across more than a full year, with the work of Chap. 3 being conducted in the winter and spring of 2014, the work of Chap. 4 being conducted in the fall of 2014 and spring of 2015, and the work of Chap. 5 being conducted in the spring of 2015. Through these temporally and spatially distinct studies, the results provide varying answers and conclusions about this type of research. With this in mind, the book serves as a comprehensive methodological guide on how to collect, process, analyze, and interpret Twitter data and offers the first frank and honest assessment of the strengths and weaknesses of Twitter data.

Social media data sets, including data from the microblogging platform Twitter, constitute a part of what is known as "Big Data," a term first used by NASA in 1997 to describe quantities of data so large that they taxed memory and hard disk capacity (Cox and Ellsworth 1997). Since around 2008, the term has been used to describe a compelling phenomenon in academia, government, business, and the media in which data is no longer seen as an entity with limited value after an initial use; rather, it is an input to be used continually in innovation, service creation, and as a means of collecting information not previously available (Mayer-Schönberger and Cukier 2013).

In the past, massive amounts of electronic information, including personally volunteered data, were not nearly as readily available as they are today. Computing capabilities were comparatively limited, and data generated was static with specific and restricted significance. In contrast, Big Data is dynamic, continually relevant, and additive (and added to) and provides unique insights and opportunities not previously available (Mayer-Schönberger and Cukier 2013). This enables researchers like ourselves to test research questions that we would not have been able to meaningfully investigate even five years ago—at least not from the angles or with the input of millions of people that we can now.

There has been considerable recent attention in the academy regarding the potential applications and pitfalls of Big Data in efforts to better understand the social world (see *Chronicle of Higher Education* special issue in April 2013). This interest is, in part, a response to the difficulties associated with collecting attitudinal data using traditional survey-based approaches, which are prohibitively costly to conduct on a wide scale and which increasingly suffer from respondent fatigue and skepticism. Data collected through publically posted online social media platforms is far cheaper to collect, less obtrusive, and much more voluminous. For example, half a billion Twitter messages are sent every day. Twitter itself is the largest microblogging platform in the world, a type of instant message service that restricts users to public posts of fewer than 140 characters. Other microblogging platforms include Sina Weibo in China and ImaHima in Japan.

Thus far, the main focus among academics has been developing new analytical tools and methodologies for efficiently processing and making sense of massive volumes of real-time information and using these tools to better understand virtual social networks. Applied researchers, planners, and policy makers are only just beginning to explore the potential of Big Data to help clarify social attitudes and potentially inform local policy and development decisions.

As of yet, few have studied whether and how Twitter posts can be used to better understand people's perception of place—that is, how they actually feel about the communities in which they live, work, and play.[1] In the business world, firms have been doing this social media "listening" for years; they call it social listening (Hinchcliffe and Kim 2012; Rappaport 2011). In this book, we push forward a new research agenda that advances knowledge and methods around urban social listening.

Why does it matter what people's perceptions of places are? Among other reasons, urban social listening can help policy makers and planners understand the overall sentiments and well-being of people living in urban areas. This is critically important particularly because the redevelopment potential of many post-industrial cities has long been stymied by the negative perceptions of investors and potential new residents and businesses. Furthermore, strong place attachment can be a galvanizing force behind community renewal efforts, which are sustained through the dedication, will, and sweat of residents who care deeply about their communities.

Researchers have begun exploring, for example, how they might determine the subjective well-being (SWB, the term used as a stand-in for happiness in psychological literature) of individuals based upon Facebook status updates (Kim and Lee 2011); how "tweets" sent by users of Twitter might be used in assisting with emergency preparedness efforts for natural disasters, epidemics, and social uprisings (Merchant et al. 2011); and how tweets provide valuable land use information for urban planners (Frias-Martinez and Frias-Martinez 2014).

In making use of the large quantities of microblogging data now available to researchers, this book will examine the sentiments residents have in places they occupy. Two central arguments underlie the book: (1) Just as social media has revolutionized social life, social listening can revolutionize the way that social scientists study cities and (2) that microblogging data is a rich data source with which to commence this new wave of urban research.

We will present in the following chapters the idea that, with the era of Big Data upon us, the study of cities will never be the same as it has been in the past. Attention to what millions of ordinary citizens are saying, within confined and narrow geographies, can provide more valid and reliable results than the obtrusive measures that have characterized social science research for more than a century. Likewise, microblogging data offers an abundant data source with which to do that listening.

While there have certainly been some naysayers (Goodspeed 2013) who question the validity of studies using these unobtrusive data sources, many others have adopted the new medium with aplomb. Some researchers have employed social network analysis to explore the ways in which individuals interact with one another (Hansen et al. 2009; Ediger et al. 2010; Catanese et al. 2010) or the ways in which people follow links (Namata et al. 2010), or combing content with user comments (Eugene Agichtein et al. 2008). Emoticons have also been analyzed to help consumers, marketers, and organizations use sentiment analysis to research products or services and analyze corresponding customer satisfaction (Go et al. 2009). Much of this research uses social media to understand group processes and properties (Tang and Lui 2010) but does little to fundamentally reveal what people think about places.

Important social science research has used massive social media data sets to advance social objectives (Ediger et al. 2010), to forecast shifts in the mood of users (Servi and Elson 2012), to enhance journalistic investigations (Diakopoulos and Shamma 2010), and to infer users' locations from their tweets (Mahmud et al. 2012). For this book, we follow a tradition of using social media to conduct opinion mining and sentiment analysis (Gokulakrishnan et al. 2012; Martineau and Finin 2009; Meeyoung Cha et al. 2010).

Although many studies have been conducted using Big Data, including some that make an effort to gauge the emotional states of users, very little work has attempted to incorporate psychological theory to interpret exactly what form of emotional well-being microblogging data measures. Moreover, there are many differing methodologies that have been used to decipher exactly what the data is telling us, with, as would be expected, the particular method chosen for a given study being related to the information that researchers are trying to infer. Before diving into a deeper literature review and the primary research of this book, we now briefly cover what the psychological literature can tell us about microblogging data, as well as provide a quick, top-level description of the methods that have been used in microblogging sentiment analysis thus far.

## Subjective Well-Being

The study and description of happiness is at least as old as the Ancient Greeks. Aristotle's defining of *eudaimonia*, often translated as happiness or welfare, is typically seen in the literature as a foundational moment, which carried through the Western philosophical cannon in the works of thinkers including Aquinas, Mill, and Bentham (Ryff and Singer 2008; Diener et al. 1998). More recently have come the humanist psychologists such as Maslow, Rogers, and Fromm and ultimately the positive psychologists who aim to change the very focus of psychology from one based solely upon the eradication of mental illness to a field encompassing the improvement of otherwise normal lives (Boniwell 2008; Sheldon and Kasser 2001).

In psychological literature, happiness is referred to as "subjective well-being," which is the term we will use throughout this section and elsewhere in this book. Subjective well-being is commonly divided into two parts: (1) life satisfaction and (2) affect, both pleasant and unpleasant (Diener et al. 1999; Boniwell 2008). Life satisfaction is an assessment an individual makes about his or her own life on the whole. Affect, on the other hand, refers to positive and negative emotions and moods as a result of events occurring in our daily lives (Diener et al. 1999; Boniwell 2008).

While much research remains to be done in teasing out the various components of both pieces of subjective well-being, psychologists have reached near consensus on the so-called Big Five personality traits that describe individual personalities in a consistent and non-overlapping way (Soldz and Vaillant 1999). These traits include neuroticism, extraversion, openness to experience, agreeableness, and conscientiousness and are referred to as NEO PI-R in the literature (n, e, and o standing for the first three traits in this list, which were the first three discovered, and PI-R standing for personality inventory—revised), demonstrating remarkable replicability across ages and cultures (McCrae and Terracciano 2005). A 45-year study of men graduating from Harvard College that followed the men throughout their life found that neuroticism and extraversion demonstrated the most correlation with life course variables such as early adult adjustment, maximum income, substance abuse, and depression (Soldz and Vaillant 1999).

In fact, the prevalence of these two traits in effecting one's subjective well-being, as revealed in personal surveys, reports of friends and acquaintances, studies of twins, daily diaries, online reports, and personality tests, has led researchers to state, "happiness is a thing called stable extraversion" (Schimmack et al. 2002; quote from Francis 1999). This form of happiness is most directly getting at the portion of subjective well-being described above as affect—that is, the difference between the pleasant moods and emotions one feels and the unpleasant moods and emotions one feels, although there is little doubt that affect is a significant contributor to the judgments one makes in rating one's life satisfaction (Davern et al. 2007).

In other words, while neuroticism and extraversion are key to affect, also known in the literature as "hedonic balance," other personality traits influence an individual's hedonic balance, and both environmental factors and hedonic balance play a significant role in determining one's overall subjective well-being. For example, chronically available sources of external information, such as one's academic or work performance or one's romantic satisfaction, play a separate role in determining life satisfaction that is uninfluenced by personality factors that create hedonic balance (Schimmack et al. 2002). In addition, evidence strongly suggests personality traits predict hedonic balance almost uniformly across cultures, while culture itself plays a moderating role between personality and life satisfaction, as hedonic balance has a greater influence in determining life satisfaction in individualistic cultures than in collectivist cultures (Schimmack

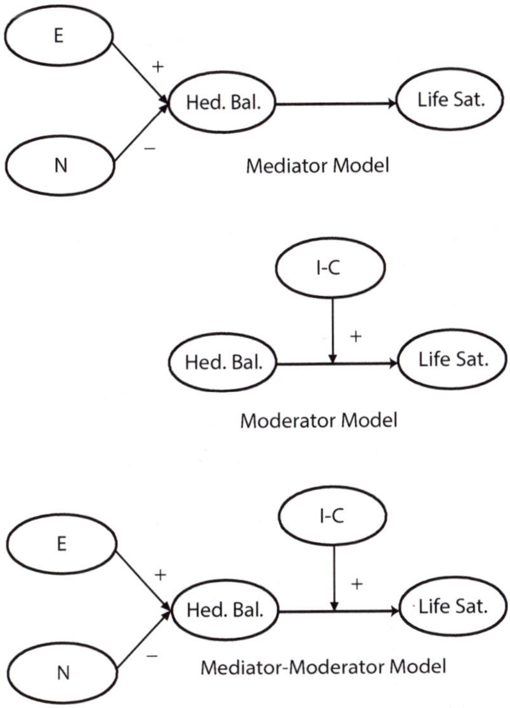


**Fig. 1.1** Models of personality, culture, and subjective well-being (Source: Schimmack et al. 2002)

et al. 2002). As shown in Fig. 1.1, this is known as the mediator-moderator model, combining the mediation of personality traits and their effect on life satisfaction indirectly via hedonic balance with the moderating influence of culture.

## Sentiment Analysis

The relationship between affect and life satisfaction is therefore an "intricate" one that does not lend itself to easy disambiguation (E. Diener, personal communication, March 16, 2015). However, a form of determining affect in particular, known as sentiment analysis (or opinion mining), has exploded in the twenty-first century as information technology has been set to the task of determining the positive and negative attitudes and opinions that people express about products, services, political campaigns, and virtually anything else that can be described in words (Pang and Lee 2008). This book relies on the sentiment analysis method of measuring subjective well-being, totaling the collective sentiments of Twitter users at the municipal level to determine the relative subjective well-being of the residents of each municipality in terms of pleasant and unpleasant affect.

To perform the sentiment analysis, a body of text must be acquired, and there must be a way of evaluating the textual body to determine the sentiments contained therein. For example, hundreds of online movie reviews have been compiled and combed for words expressing positive or negative sentiment (Pang et al. 2002). This study used multiple machine learning methods in which the researchers attempted to train the analysis software to make it more accurate in classifying sentiments. Prabowo and Thelwall (2009) tested three different forms of sentiment analysis on movie reviews, product reviews, and MySpace comments—rule-based classification, supervised learning, and machine learning—concluding that a hybrid approach produced the best results. Each of these approaches requires that researchers actively participate in the creation of rules and algorithms to be used in determining sentiment classification.

As these studies demonstrate, sentiment analyses can easily become rather complex, involving multiple iterations of testing and improving upon a classification scheme. They may also involve the classification of single words, phrases, sentences, or entire documents, depending on the objective of the study. Nasukawa and Yi (2003) used the subject(s) and object(s) of a sentence as the entities upon which sentiment is applied, finding this method to be far more accurate than the classification of entire documents. A case in point is that a movie review may have multiple positive and negative statements, sometimes even within one sentence, and negative descriptions of, for example, a specific actor or the film's cinematography may contrast with an overall positive review of the film (Nasukawa and Yi 2003). In these more complicated research cases, the objective of a sentiment analysis is to determine the "contextual polarity" of a given sentiment; that is, given the context of the content being analyzed, is the sentiment expressed positive or negative (Wilson et al. 2009)?

A simpler approach, which does not make contextual distinctions on the fly and instead relies on the characteristic polarity of a word, is the lexicon-based approach. In this methodology, researchers use a dictionary containing thousands of sentiment-expressing words that have been determined to be either positive or negative, with rigorous testing of the dictionary to ensure validity being clearly preferable (Taboada et al. 2011). The words within a given textual body may be summed into totals of positive and negative word uses or may be given a score meant to capture the degree of sentiment for particular words—for instance, on a range from −3 to +3. In this latter methodology, "good" might be given a score of +1, "bad" a score of −1, "great" a score of +2, and "amazing" a score of +3. This book uses the latter approach, multiplying each sentiment-containing word by its associated score and then summing the total scores of positive and negative words across all tweets.

A primary limitation of this particular form of sentiment analysis is that a statement with an obviously negative sentiment to human eyes may not contain a specific keyword contained in the dictionary being used by the sentiment analysis software that would be classified as negative (Pang et al. 2002). For instance, if someone were to review the restaurant she ate lunch at today with the question, "Who could eat this stuff?," it would not generate a negative score in a sentiment classifier that simply looks for negative terms. Alternatively, a word that is typically classified as explicitly positive or negative may not actually be expressing a sentiment (Wilson et al. 2009). An example of this would arise when talking about a land trust, with the word "trust" being classified as expressing a positive sentiment when it is in fact being used as a noun. While the law of large numbers would imply that the averaged values of the sentiment of individual word would converge to the average of the emotions for the sample, the ambiguity of language and other limitations will be discussed further throughout the remainder of the book.

## Outline of Chapters

In the next chapter, we provide some historical context for social listening and urban policy, reviewing key areas of the literature. In Chap. 3, we explore what this kind of social listening looks like in a small post-industrial city in New England by looking closely at the sentiment of tweets and keywords employed. The successes and failures in this test-drive help generate some of the methods and approaches we then apply in the next two chapters and recommend for future researchers.

In Chap. 4, we begin by posing a more focused question than in the previous chapter, exploring the ways that urban immigrants express themselves on Twitter differently than native English speakers do. By comparing Portuguese tweets in several immigrant gateway cities with demographic data from the US Census, the chapter examines the challenges and pitfalls of a comparative language study of microblogging data.

Chapter 5 offers a broader, national view of eight US cities where Twitter data was collected in 2013 and 2015, employing a more rigorous statistical analysis of the differences between Twitter data and more conventional census sources. The results suggest that microblogging data has particular strengths and weaknesses of which researchers must be aware. The successes of the data collection and analysis in this chapter serve as a blueprint for the recommended methodological steps outlined in the final chapter.

Chapter 6 concludes the book by synthesizing the findings presented, offering implications for both urban studies and planning, as well as future research that endeavors to make use of social listening methods. The book ends by recommending a series of concrete steps for approaching urban social listening that draws on both theory and practice presented in the previous chapters.

## Note

1. A notable exception is Schweitzer (2014).

## References

Agichtein, E., Castillo, C., Donato, D., Gionis, A., & Mishne, G. (2008). Finding high-quality content in social media. In *Proceedings of the 2008 International Conference on Web Search and Data Mining*, ACM, 183–194.

Boniwell, I. (2008). *Positive psychology in a nutshell: A balanced introduction to the science of optimal functioning*. London: Personal Well-Being Centre.

Catanese, S., De Meo, P., Ferrara, E., & Fiumara, G. (2010). Analyzing the Facebook friendship graph. arXiv preprint arXiv:1011.5168.

Cha, M., Haddadi, H., Benevenuto, F., & Gummadi, K. P. (2010). Measuring user influence in Twitter: The Million Follower Fallacy. In *Proceedings of the Fourth Annual AAAI Conference on Weblogs and Social Media*, 10–17.

Cox, M., & Ellsworth, D. (1997). Application-controlled demand paging for out-of-core visualization. In *Proceedings of the 8th Conference on Visualization*, IEEE Computer Society Press, 235-ff.

Davern, M., Cummins, R., & Stokes, M. (2007). Subjective wellbeing as an affective-cognitive construct. *Journal of Happiness Studies, 8*(4), 429–449.

Diakopoulos, N. A., & Shamma, D. A. (2010). Characterizing debate performance via aggregated twitter sentiment. *Proceedings of the SIGCHI Conference on Human Factors in Computing Systems,* ACM.

Diener, E. (2015, March 16). Personal communication.

Diener, E., Sapyta, J. J., & Suh, E. (1998). Subjective well-being is essential to well-being. *Psychological Inquiry, 9,* 33–37.

Diener, E., Suh, E. M., Lucas, R. E., & Smith, H. E. (1999). Subjective well-being: Three decades of progress. *Psychological Bulletin, 125,* 276–302.

Ediger, D., Jiang, K., Riedy, J., Bader, D. A., Corley, C., Farber, R., & Reynolds, W. N. (2010). Massive social network analysis: Mining Twitter for social good. *39th International Conference on Parallel Processing,* 583–593.

Francis, L. J. (1999). Happiness is a thing called stable extraversion: A further examination of the relationship between the Oxford Happiness Inventory and Eysenck's dimensional model of personality and gender. *Personality and Individual Differences, 26*(1), 5–11.

Frias-Martinez, V., & Frias-Martinez, E. (2014). Crowdsourcing land use maps via Twitter. *KDD'14, New York.*

Go, A., Bhayani, R., & Huang, L. (2009). Twitter sentiment classification using distant supervision. *CS224N Project Report,* Stanford, 1–12.

Gokulakrishnan, B., Priyanthan, P., Ragavan, T., Prasath, N., & Perera, A. (2012). Opinion mining and sentiment analysis on a Twitter data stream. *The International Conference on Advances in ICT for Emerging Regions—ICTer 2012,* 182–188.

Goodspeed, R. (2013). The limited usefulness of social media and digital trace data for urban social research, *AAAI Technical Report,* 2–4.

Hansen, D. L., Rotman, D., Bonsignore, E., Milic-Frayling, N., Rodrigues, E. M., Smith, M., & Shneiderman, B. (2009). *Do you know the way to SNA?: A process model for analyzing and visualizing social media data* (pp. 1–10). College Park: University of Maryland.

Hinchcliffe, D., & Kim, P. (2012). Social business by design: Transformative social media strategies for the connected company. Wiley.

Kim, J., & Lee, J.-E. R. (2011). The Facebook paths to happiness: Effects of the number of Facebook friends and self-presentation on subjective well-being. *Cyberpsychology, Behavior and Social Networking, 14*(6), 359–364.

Mahmud, J., Nichols, J., & Drews, C. (2012). Where is this tweet from? Inferring home locations of Twitter users. In *Proceedings of the Sixth International AAAI Conference on Weblogs and Social Media,* 511–514.

Martineau, J., & Finin, T. (2009). Delta TFIDF: An improved feature space for sentiment analysis. In *Proceedings of the Third International ICWSM Conference,* 258–261.

Mayer-Schönberger, V., & Cukier, K. (2013). *Big data: A revolution that will transform how we live, work, and think.* Boston: Eamon Dolan/Houghton Mifflin Harcourt.

McCrae, R. R., & Terracciano, A. (2005). Universal features of personality traits from the observer's perspective: Data from 50 cultures. *Personality Processes and Individual Differences, 88*(3), 547–561.

Merchant, R. M., Elmer, S., & Lurie, N. (2011). Integrating social media into emergency-preparedness efforts. *New England Journal of Medicine, 365*(4), 289–291.

Namata, G. M., Sharara, H., & Getoor, L. (2010). A survey of link mining tasks for analyzing noisy and incomplete networks. Springer.

Nasukawa, T., & Yi, J. (2003). *Sentiment analysis: Capturing favorability using natural language processing* (pp. 70–77). New York: ACM.

Pang, B., & Lee, L. (2008). Opinion mining and sentiment analysis. *Foundations and Trends in Information Retrieval., 2*(1–2), 1–90.

Pang, B., Lee, L., & Vaithyanathan, S. (2002). Thumbs up? Sentiment classification using machine learning techniques. *EMNLP*, 79–86.

Prabowo, R., & Thelwall, M. (2009). Sentiment analysis: A combined approach. *Journal of Informetrics, 3*(1), 143–157.

Rappaport, S. D. (2011). *Listen first!: Turning social media conversations into business advantage.* New York: Wiley.

Ryff, C. D., & Singer, B. H. (2008). Know thyself and become what you are: Aeudaimonic approach to psychological well-being. *Journal of Happiness Studies, 9*(1), 13–39.

Schimmack, U., Diener, E., & Oishi, S. (2002). Life-satisfaction is a momentary judgment and a stable personality characteristic: The use of chronically accessible and stable sources. *Journal of Personality, 70*, 345–384.

Schimmack, U., Radhakrishnan, P., Oishi, S., Dzokoto, V., & Ahadi, S. (2002). Culture, personality, and subjective well-being: Integrating process models of life satisfaction. *Personality Processes and Individual Differences, 82*(4), 582–593.

Schweitzer, L. (2014). Planning and social media: A case study of public transit and stigma on Twitter. *Journal of the American Planning Association, 80*(3), 218–238.

Servi, L., & Elson, S. B. (2012). A Mathematical approach to identifying and forecasting shifts in the mood of social media users. *The MITRE Corporation*, 27-1–27-16.

Sheldon, K. M., & Kasser, T. (2001). Getting older, getting better: Personal strivings and psychological maturity across the lifespan. *Developmental Psychology, 37*, 491–501.

Soldz, S., & Vaillant, G. E. (1999). The Big five personality traits and the life course: A 45-year longitudinal study. *Journal of Research in Personality, 33*, 208–232.

Taboada, M., Brooke, J., Tofiloski, M., Voll, K., & Stede, M. (2011). Lexicon-based methods for sentiment analysis. *Computational Linguistics, 37*(2), 267–307.

Tang, L., & Liu, H. (2010). Community detection and mining in social media. *Synthesis Lectures on Data Mining and Knowledge Discovery, 2*(1), 1–137.

Wilson, T., Wiebe, J., & Hoffmann, P. (2009). Recognizing contextual polarity: An exploration of features for phrase-level sentiment analysis. *Computation Linguistics, 35*(3), 400–433.

CHAPTER 2

# A (Short) History of Social Media Sentiment Analysis

**Abstract** In this chapter, we introduce the concept of sentiment analysis as a way to study the prevalence of positive or negative sentiments in expressed attitudes and opinions. We review the ways researchers broadly have used sentiment analyses of Twitter data and other digital data and the established benefits and limits of this method for urban social science research. We then discuss how Twitter data and other similar forms of social media data have been applied in a wide variety of urban planning issues and projects across the globe.

**Keywords** Social listening • Social media • Twitter • Microblogging

In this chapter, we offer a bit more perspective on urban social listening by examining the associated data sources and tools, as well as how it has been used in the urban studies, urban affairs, and urban planning literatures.

## Microblogging Sentiment Analysis

As Twitter posts are the form of Big Data discussed in this book, this section will examine the ways in which sentiment analyses—qualitative measurements of opinions and attitudes—of Twitter data have been used for a myriad of academic research purposes. Furthermore, we will discuss

© The Editor(s) (if applicable) and The Author(s) 2016

J.B. Hollander et al., *Urban Social Listening*,
DOI 10.1057/978-1-137-59491-4_2

how Twitter data and other similar forms of social media data have been applied in a wide variety of urban planning contexts across the globe.

Sentiment analysis, sometimes referred to as opinion mining, is the analysis of expressed attitudes and opinions with the goal of determining the degree to which there are positive or negative sentiments therein (Liu 2012). In this qualitative method, a score is assigned to each textual entity within a larger body of data based upon the strength of modifying words within that entity (Godbole et al. 2007). For example, a tweet containing the word "good" would be assigned a positive score, a tweet containing the words "very good" would be assigned an even higher positive score, a tweet with the word "bad" would be assigned a negative score, and so on.

As would be expected, there are both beneficial and limiting features of microblogging sentiment analyses that must be kept in mind when conducting such research. On the beneficial side are its relative ease and low cost, with people who use Twitter (or Facebook, or Foursquare, or any number of other platforms) freely volunteering information on a virtually unlimited number of topics. Traditional surveys or polls are more time intensive, cost far more to conduct, and can cover only so many topics. On the other hand, microblogging applications have not been universally adopted by the general public and skew toward a younger demographic (Gayo-Avello 2011). There may be other demographic factors at play, such as socioeconomic status, meaning that a representative sample of the broader population is unlikely. Additionally, language is complex, contextual, and ever changing. Machines, for all their processing power, are sometimes incapable of interpreting linguistic subtleties.

Nonetheless, there are multiple ways in which microblogging data may effectively influence or even lead to the creation of policies, programs, and institutions. Particularly relevant to our work are studies that investigate different measures of happiness as calculated based on tweets. The results of prior research generally support the findings from the psychology literature reviewed in the first chapter of this book. For example, Bliss et al. (2012) studied sentiment among Twitter users in common social networks, as indicated by replying to one another's tweets (Bliss et al. 2012). They found that happiness is assortative, although the researchers did not attempt to distinguish whether the association was due primarily to homophily (birds of a feather flock together) or contagion (happiness is infectious). These results are supported by a similar study by Bollen et al. (2011b), who found that people with relatively higher or lower degrees of subjective well-being connect with others of similar subjective well-being, suggesting homophily (Bollen et al. 2011b).

Microblogging has also been shown to be a fairly reliable indicator of swings in public sentiment. Bollen et al. (2011a) compare sentiment analysis of daily tweets over a five-month period in 2008 to the timing of newsworthy events. They find that social, political, cultural, and economic events are correlated with significant changes in public mood as represented by tweets (Bollen et al. 2011a). A lengthier study of nearly 4.6 billion tweets over 33 months similarly demonstrated a strong relationship between notable events in the news and the sentiments expressed about the participants in those events (Dodds et al. 2011).

Some have even gone so far as to use Twitter to predict the outcomes of important events. One study found that a high level of emotion, positive or negative, in tweets on a given day correlated with decreases in stock market indexes the following day, while a low level of emotional tweeting correlated with an increase in stock market indexes the following day (Zhang et al. 2011b). The same authors analyzed retweets in the USA containing the words "hope," "fear," or "worry," as well as certain economic keywords like "dollar," "gold," or "oil," and found there to be significant correlations between market movement and retweets for most, but not all, of the keywords (Zhang et al. 2011a).

A particularly lively debate has centered on the potential ability of Twitter to predict elections. Tumasjan et al. (2010) conducted a count and sentiment analysis of tweets leading up to the German federal election of 2009. They found that the volume of tweets mentioning each political party closely corresponded to the ranking of those parties in terms of vote share. The researchers were surprised by this result given that Twitter's sample is likely unrepresentative of the voting public. They did note that Twitter users tend to be more highly educated than the public as a whole and have more influence on media pundits who frame the political debate for a nationwide audience, though no insights were offered in the study about actual voting behavior (Tumasjan et al. 2010). Jungherr et al. (2012) directly rebutted those arguments by pointing out that Tumasjan et al. (2010) neglected to include the Pirate Party, which had almost twice as many mentions in tweets leading up to the German federal election in 2009 as the next closest party (the Christian Democrats, who actually received the most votes). Sang and Bos (2012) used counts of political parties mentioned in tweets to analyze results of the 2012 Dutch senate elections, finding that these counts would have accurately predicted the results for 52 out of 75 seats, a figure that was improved to 57 out of 75 seats when a sentiment analysis was introduced. Other researchers, however, have echoed the words of Jungherr et al. (2012), stating that no elections have actually been predicted, no common method of establishing Twitter "votes" has been agreed upon, and there is no clear comparison between tweets and other data sources like polls, election results, or party shares of seats (Metaxas and Mustafaraj 2012; Gayo-Avello 2012).

O'Connor et al. (2010) suggested that tweets might actually be getting better at predicting events over time—presumably as the platform gains more active and intense users. They found an increase in the degree of correlation between sentiments for tweets containing the words "economy," "job," or "jobs" with traditional measures of consumer confidence. They also reported stronger correlations between tweet sentiments about Obama and his job approval rating in 2009 than in public polling during 2008. On the other hand, the frequency of tweets was found to have a stronger correlation with polls than sentiment scores (O'Connor et al. 2010).

Twitter has also been used to study national election results at subnational level; however, no one, to our knowledge, has yet conducted an in-depth investigation into how well this platform predicts the outcomes of local issues, which are often of particular interest to planners and municipal policy makers. Of particular note is a recent study by Gordon (2013), who analyzed state results in the 2008 and 2012 US presidential elections. He found that the volume of tweets mentioning a particular party or candidate has a pro-Democratic Party bias (likely because of a liberal-skewing Twitter user base). The predictive accuracy of tweets improved significantly between 2008 and 2012, although this may be due to the latter being a more closely contested election. This research also found that a sentiment analysis of tweets that assigned a "vote" for Mitt Romney or Barack Obama to each Twitter user produced a more accurate picture of election results (at the state level) than a simple volume count of tweet mentions, although there was still a pro-Obama bias (Gordon 2013).

In an attempt to use Twitter to aid public health planning, a study by Eichstaedt et al. (2015) established that an analysis of language patterns of tweets, using preestablished dictionaries for positive and negative emotions, positive and negative social relationships, and engagement and disengagement, was a better predictor of heart disease mortality at the county level than a model of common demographic, socioeconomic, and health risk factors, such as smoking and diabetes.

## Microblogging and Comparative Language Studies

Several studies have used microblogging data to make comparisons between linguistic and cultural groups so as to facilitate an improved understanding of multicultural communities (Hale 2014; Magdy et al. 2014; Mocanu et al. 2013; Wilkinson and Thelwall 2012). Hale (2014) focused on Twitter users who tweet in more than one language to show how these users are instrumental in bridging language gaps and disseminating information from users of one language to users of another. Madgy et al. (2014) analyzed geotagged tweets from a number of countries around the world in order to compare linguistic diversity among different countries. Their unit of comparison is the country scale and, rather than analyzing the content of each tweet, they simply used the language tag associated with each Twitter user.

In a similar study, Mocanu et al. (2013) used language data associated with geotagged tweets to map language usage on Twitter both within and between countries. Wilkinson and Thelwell (2012), on the other hand, focused only on English but compared how different topics were emphasized in different international and cultural contexts. Cui et al. (2011) discussed the methodology of using emoticons to detect sentiment as way to bridge the gap between tweets composed in different languages. Gao et al. (2012) compared sentiment analysis of microblogging data inside and outside of China by comparing results from Twitter and from Sina Weibo, the most popular Chinese microblogging service. Poblete et al. (2011) compared Twitter sentiment and languages used in the top ten countries for Twitter usage worldwide.

## Microblogging and Urban Studies

The use of Twitter and other microblogging data to enhance urban planning and related studies of place is a relatively new development due to the newness of the platform itself—Twitter was only launched in the summer of 2006. Despite Twitter still being less than a decade old, numerous efforts have been made, both inside and outside of academia, to apply analysis of a corpus of tweets from particular locations to a variety of urban issues and planning endeavors.

Several studies use Twitter-based sentiment analysis to understand what is happening in urban places through examinations of the geography of well-being and happiness. A 2013 study by Mitchell et al. used tweets to calculate the happiest and saddest cities and states. They also showed that various factors such as frequency of tweets, use of obscenities, and obesity rates have consistent correlations with places showing less happiness (Mitchell et al. 2013). An analysis of neighborhoods in London showed a strong link between sentiments expressed in tweets emanating from those neighborhoods and traditional measures of that community's socioeconomic well-being, such as income and crime (Quercia et al. 2012). These kinds of investigations into the relationship between particular places and the mood of tweets have also been applied to specific places and events, such as the 2012 London Olympic Games; 2013 Milan Design Week (Balduini et al. 2013); the city of Newcastle upon Tyne (Mearns et al. 2014); 15 museums in Yorkshire County, UK (Lovelace et al. 2014); and the whole of New York City (Bertrand et al. 2013). These studies add to the growing body of evidence that, through the analysis of tweets, one can learn about sentiment in particular places.

On the practitioner side, planners in Brisbane, Australia, used a system they called Discussions in Space (DIS), in which a large public screen encouraged residents to participate, via Twitter, in the city's long-term visioning process (Schroeter and Houghton 2011). Residents were asked to send in their "bright ideas" for the future of Brisbane, either by sending a text to a specific number or by using an exclusive hashtag, with several tweets deemed to be relevant or interesting posted to the public screen. An additional study focusing on suburban areas of Brisbane asked participating mothers to check in via one of three mobile phone applications at every location where they brought their children for physical activity over a one-week period, with implications for improving public health (Ben-Harush et al. 2012).

As might be expected, many urban-focused uses of Twitter data involve mapping where tweets are coming from and analyzing the content of those tweets. For example, MacEachren et al. (2011) created a web-based application to query the Twitter API for tweets relevant to crisis management and disaster relief efforts, accommodating not only the geolocation of the tweets but locations mentioned in the contents of the tweets. The latter is particularly important for emerging planning applications, as many Twitter users do not enable the location tracking options of Twitter. Sakaki et al. (2010) established the ability of their tweet classifier to detect earthquakes and send out warning emails faster than the Japan Meteorological Association's Broadcasts. Antonelli et al. (2014) created a program with a dashboard of information, including maps and timelines, displaying the location of tweets during citywide events, with an eye toward ultimately including sentiment analysis as part of the dashboard's reports. A similar paper describes a dashboard of tweet location data coming from the 2012 London Olympic Games and 2013 Milan Design Week, including a sentiment analysis of the latter (Balduini et al. 2013).

Other city-level studies include the creation of clusters of Foursquare[1] check-ins posted to Twitter in several neighborhoods of Pittsburgh, enabling the creation of maps of behavior patterns for clusters of residents that do not conform to traditional neighborhood boundaries (Cranshaw et al. 2012). A similar study investigated public mood via sentiment analysis of tweets in New York City, showing how attitudes shifted in relation to nearby landmarks or facilities such as Times Square, hospitals, and jails (Bertrand et al. 2013). A smaller scale was adopted in a study examining visitor activity at 15 museums in Yorkshire, UK, from nearby residential areas (Lovelace et al. 2014).

Researchers have also been using Twitter to study the movement of people, providing new insights in the realm of traffic analysis and changing land use/activity patterns. Fujisaka et al. (2010) focused on the aggregation and dispersion of people in different parts of Tokyo as determined by geolocated tweets. In a study of geolocated tweets in New York, London, and Madrid, Frias-Martinez et al. (2013) showed how land use can be identified by changes in tweet volume throughout the day. Similarly, Wakamiya et al. (2011) used geotagged tweets to classify Japanese municipalities into one of four categories—bedroom, office, nightlife, and multifunctional—based upon number of tweets, number of Twitter users, and movement of Twitter users. Mobility patterns have studies with social media applications that require users to "check in" to locations they visit (Cheng et al. 2011).

Many municipal agencies and regional authorities have begun crafting their own social media outreach wings in an effort to engage the citizenry and, at least informally, assess public sentiment (Evans-Cowley and Hollander 2010). Given the relatively low cost of such efforts, we expect more and more cities to turn to social media. However, little work has been done to evaluate the accuracy or value of microblogging tracking and sentiment analysis with regard to issues of local concern, such as attitudes over development impacts, service provision, effective governance, or predictions of local referenda and elections. A recent study of tweets at the local level provided conclusive evidence that transit agencies engaging more actively with other Twitter users, as opposed to simply blasting out information without the potential for a dynamic dialogue, experienced a significantly improved level of Twitter discourse surrounding public transit (Schweitzer 2014).

## Conclusion

In this brief history of the use of microblogging data in social science research and urban studies, we have shown that there is a great deal of formative research and real potential for this approach. Not all of these examples deliver on the promise of a new technology, but what they offer us collectively is a framework for how to begin to test whether and how Twitter data can be used to produce valid and reliable studies of cities. What follows in the next three chapters are the results of three efforts we undertook to build on the literature cited above and to participate in urban social listening.

## Note

1. Foursquare is a mobile app that provides place recommendations based on the user's identity and current location.

## References

Antonelli, F., Azzi, M., Balduini, M., Ciuccarelli, P., Valle, E. D., & Larcher, R. (2014, May). City sensing: Visualising mobile and social data about a city scale event. In *Proceedings of the 2014 International Working Conference on Advanced Visual Interfaces*, ACM, 337–338.

Balduini, M., Della Valle, E., Dell'Aglio, D., Tsytsarau, M., Palpanas, T., & Confalonieri, C. (2013). Social listening of city scale events using the streaming linked data framework. In *The semantic web–ISWC 2013* (pp. 1–16). Berlin/Heidelberg: Springer.

Ben-Harush, O., Carroll, J.-A., & Marsh, B. (2012). Using mobile social media and GIS in health and place research. *Continuum, 26*(5), 715–730.

Bertrand, K. Z., Bialik, M., Virdee, K., Gros, A., & Bar-Yam, Y. (2013, August 20). Sentiment in New York City: A high resolution spatial and temporal view. arXiv:1308.5010.

Bliss, C. A., Kloumann, I. M., Harris, K. D., Danforth, C. M., & Dodds, P. S. (2012). Twitter reciprocal reply networks exhibit assortativity with respect to happiness. *Journal of Computational Science, 3*(5), 388–397.

Bollen, J., Mao, H., & Pepe, A. (2011a). Modeling public mood and emotion: Twitter sentiment and socio-economic phenomena. In *Proceedings of the Fifth International AAAI Conference on Weblogs and Social Media*, 450–453.

Bollen, J., Gonçalves, B., Ruan, G., & Mao, H. (2011b). Happiness is assortative in online social networks. *Artificial Life, 17*(3), 237–251.

Cheng, Z., Caverlee, J., Lee, K., & Sui, D. Z. (2011). Exploring millions of footprints in location sharing services. *ICWSM*, Barcelona, Spain, 17 to 21 July 2011, pp. 81–88.

Cranshaw, J., Schwartz, R., Hong, J., & Sadeh, N. (2012). The livehoods project: Utilizing social media to understand the dynamics of a city. *Sixth International AAAI Conference on Weblogs and Social Media*, 58–65.

Cui, A., Zhang, M., Liu, Y., & Ma, S. (2011). Emotion tokens: Bridging the gap among multilingual twitter sentiment analysis. *7th Asia Information Retrieval Societies Conference*, AIRS.

Dodds, P. S., Harris, K. D., Kloumann, I. M., Bliss, C. A., & Danforth, C. M. (2011). Temporal patterns of happiness and information in a global social network: Hedonometrics and Twitter. *PloS One, 6*(12), e26752. doi:10.1371/journal.pone.0026752.

Eichstaedt, J. C., et al. (2015). Psychological language on Twitter predicts county-level heart disease mortality. *Psychological Science, 26*(2), 159–169.

Evans-Cowley, J., & Hollander, J. (2010). The new generation of public participation: Internet-based participation tools. *Planning, Practice & Research., 25*(3), 397–408.

Frias-Martinez, V., Soto, V., Hohwald, H., & Frias-Martinez, E. (2013). Sensing urban land use with Twitter activity. http://citeseerx.ist.psu.edu/viewdoc/download?doi=10.1.1.409.878&rep=rep1&type=pdf

Fujisaka, T., Lee, R., & Sumiya, K. (2010). Exploring urban characteristics using movement history of mass mobile microbloggers. In *Proceedings of the Eleventh Workshop on Mobile Computing Systems & Applications*, ACM.

Gao, Q., Abel, F., Houben, G. J., & Yu, Y. (2012). A comparative study of users' microblogging behavior on Sina Weibo and Twitter. In *User modeling, adaptation, and personalization* (pp. 88–101). Berlin/Heidelberg: Springer.

Gayo-Avello, D. (2011). Limits of electoral predictions using Twitter. In *Proceedings of the Fifth International AAAI Conference on Weblogs and Social Media*, AAAI, 490–493.

Gayo-Avello, D. (2012). I wanted to predict elections with Twitter and all I got was this Lousy Paper—A balanced survey on election prediction using Twitter data. arXiv:1204.6441.

Godbole, N., Srinivasaiah, M., & Skiena, S. (2007). Large-scale sentiment analysis for news and blogs. *ICWSM, 7*, 21.

Gordon, J. (2013). *Comparative geospatial analysis of Twitter sentiment data during the 2008 and 2012 US Presidential elections.* Eugene: University of Oregon.

Hale, S. A. (2014). Global connectivity and multilinguals in the Twitter network. In *Proceedings of the SIGCHI Conference on Human Factors in Computing Systems*, 833–842.

Jungherr, A., Jürgens, P., & Schoen, H. (2012). Why the Pirate Party won the German election of 2009 or the trouble with predictions: A response to Tumasjan, A., Sprenger, T. O., Sander, P. G., & Welpe, I. M. 'Predicting elections with Twitter: What 140 characters reveal about political sentiment. *Social Science Computer Review, 30*, 229–234.

Liu, B. (2012). *Sentiment analysis and opinion mining*. San Rafael: Morgan & Claypool Publishers.

Lovelace, R., Malleson, N., Harland, K., & Birkin, M. (2014). *Geotagged tweets to inform a spatial interaction model: A case study of museums*. Leeds: School of Geography, University of Leeds.

MacEachren, A. M., et al. (2011). Geo-twitter analytics: Applications in crisis management. *25th International Cartographic Conference*.

Magdy, A., Ghanem, T. M., Musleh, M., & Mokbel, M. F. (2014). Exploting geotagged tweets to understand local language diversity. *Proceedings of Workshop on Managing and Mining Enriched Geo-Spatial Data*, ACM. 2.

Mearns, G., Simmonds, R., Richardson, R., Turner, M., Watson, P., & Missier, P. (2014). Tweet my street: A cross-disciplinary collaboration for the analysis of local Twitter data. *Future Internet, 6*(2), 378–396.

Metaxas, P. T., & Mustafaraj, E. (2012). Social media and the elections. *Policy Forum, 338*, 472–473.

Mitchell, L., Harris, K. D., Frank, M. R., Dodds, P. S., & Danforth, C. M. (2013). The Geography of happiness: Connecting Twitter sentiment and expression, demographics, and objective characteristics of place. *PloS One, 8*(5), e64417.

Mocanu, D., Baronchelli, A., Perra, N., Gonçalves, B., Zhang, Q., & Vespignani, A. (2013). The Twitter of Babel: Mapping world languages through microblogging platforms. *PloS One*. doi:10.1371/journal.pone.0061981.

O'Connor, B., Balasubramanyan, R., Routledge, B. R., & Smith, N. A. (2010). From tweets to polls: Linking text sentiment to public opinion time series. *ICWSM, 11*, 122–129.

Poblete, B., Garcia, R., Mendoza, M., & Jaimes, A. (2011). Do all birds tweet the same? Characterizing Twitter around the world. In *Proceedings of the 20th ACM international conference on information and knowledge management*, ACM, 1025–1030.

Quercia, D., Ellis, J., Capra, L., & Crowcroft, J. (2012). Tracking gross community happiness from tweets. *15th ACM Conference on Computer Supported Cooperative Work (CSCW)*. Seattle.

Sakaki, T., Okazaki, M., & Matsuo, Y. (2010, April 26–30). Earthquake shakes Twitter users: Real-time event detection by social sensors. *WWW2010*, Raleigh.

Sang, E. T. K., & Bos, J. (2012). Predicting the 2011 Dutch senate election results with Twitter. In *Proceedings of the Workshop on Semantic Analysis in Social Media*, 53–60.

Schroeter, R., & Houghton, K. (2011). Neo-planning: Location-based social media to engage Australia's new digital locals. *Proceedings of Planning Institute of Australia National Conference 2011*. Planning Institute of Australia.

Schweitzer, L. (2014). Planning and social media: A case study of public transit and stigma on Twitter. *Journal of the American Planning Association, 80*(3), 218–238.

Tumasjan, A., Sprenger, T. O., Sandner, P. G., & Welpe, I. M. (2010). Predicting elections with Twitter: What 140 characters reveal about political sentiment. In *Proceedings of the Fourth International AAAI Conference on Weblogs and Social Media*, AAAI, 178–184.

Wakamiya, S., Lee, R., & Sumiya, K. (2011). Crowd-based urban characterization: Extracting crowd behavioral patterns in urban areas from twitter. In *Proceedings of the 3rd ACM SIGSPATIAL* international *workshop on location-based social networks*, ACM.

Wilkinson, D., & Thelwall, M. (2012). Trending Twitter topics in English: An international comparison. *Journal of the American Society for Information Science and Technology, 63*(8), 1631–1646.

Zhang, X., Fuehres, H., & Gloor, P. A. (2011a). Predicting stock market indicators through Twitter: 'I hope it is not as bad as I fear. *Procedia – Social and Behavioral Sciences, 26*, 55–62.

Zhang, X., Fuehres, H., & Gloor, P. A. (2011b). Predicting asset value through Twitter Buzz. In J. Altmann et al. (Eds.), *Advances in collective intelligence* 2011, AISC 113, 23–34.

CHAPTER 3

# Taking Microblogging Data for a Test Drive

**Abstract** In this chapter we provide an account of our attempt to analyze Twitter data. We describe our methods for creating a database of over 100,000 tweets produced by users in the city of New Bedford, Massachusetts. We attempt to analyze the way in which the Twitter messages engaged with the topics of urban policy and find there is a cursory overlap. We also compare the commercially available IBM SPSS Modeler to our custom-designed sentiment analyzer. Both methods showed a relatively low percentage of sentiment overall but a greater prevalence of positive tweets. Overall, we note that the magnitude of microblogging data and the ability to capture it readily and improvements in analysis techniques may allow for quantity to compensate for low percentages of sentiment.

**Keywords** New Bedford • Twitter • Sentiment analysis

Inspired by scores of articles, conference presentations, and anecdotal accounts, we commenced in 2013 an effort to collect and analyze data from the Twitter API. While a few tools were available at the time, none were easy to use and most required the prospective researcher to engage in extensive computer programming. We designed a piece of software, Urban Attitudes, that stored, indexed, and analyzed a continuous stream

© The Editor(s) (if applicable) and The Author(s) 2016
J.B. Hollander et al., *Urban Social Listening*,
DOI 10.1057/978-1-137-59491-4_3

of data from the Twitter Decahose (a 10% sample of all tweets) that were geotagged to specific locations (details on the software are presented in Appendix 1).

## Exploring Strategies for Using the Twitter API

Below we chronicle our initial pursuit of an iteratively developed software intended to explore how the Twitter API could be used to address a core urban planning question: what does the public think? Conventional approaches to answering such a question have tended to involve surveys, focus groups, or interviews (Gaber and Gaber 2007). The goal for us was to offer a cursory but systematic analysis of the potential for Twitter data to shed light on the sentiments of users in a city—in this case, the city of New Bedford, Massachusetts. We selected New Bedford due to its small population (75,000), small geographic area (24 square miles), and low average age (23% under 18 years of age). In addition, we were already asking these same questions using conventional qualitative and quantitative research methods in New Bedford in which we were studying children and family policy with Professor Tama Leventhal of the Eliot-Pearson Department of Child Study and Human Development at Tufts University.

Our custom software (Urban Attitudes) pulled only those messages that were tweeted from a geographic areas delimited as a rectangle around New Bedford (see Fig. 3.1). Using 70.89941601 (longitude) and 41.759387677 (latitude) for the Northeast corner and –70.981963 (longitude) and 41.591322004 (latitude) for the Southwest corner, the rectangle encompassed the entire geographic extent of New Bedford, along with small slices of some neighboring communities (see Fig. 3.1). We collected a total of 122,187 messages during the period of February 9–April 3, 2014.

Once we had acquired and organized the data (using the custom Urban Attitude software), we proceeded to conduct a sentiment and keyword analysis. To conduct the analysis, we considered many of the software applications described in Chap. 2. We arrived upon a professional software application because it is both convenient and practical for urban planning purposes. We first set out to select existing applications. SAS and STATA both offered sentiment analysis add-on packages, though at a high cost to our institutions. We found that the open-access software R has several respectable sentiment analysis programs, but our own (and our research assistants') unfamiliarity with R made it an unattractive option. We set-

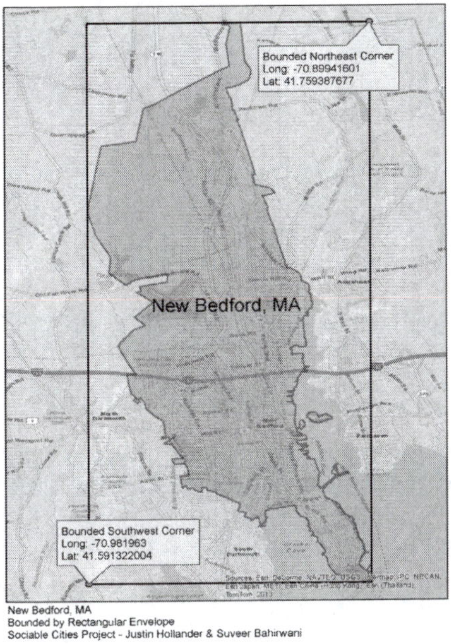


**Fig. 3.1** City of New Bedford, enveloped by a rectangle to facilitate downloading of Twitter messages

tled on IBM SPSS Modeler, which we were provided with *gratis* through a license with IBM for one year. Because Modeler did not satisfy all of our research needs, we also added a sentiment analyzer module into our Urban Attitudes software.

Using IBM SPSS Modeler, we ran the Twitter messages through a sentiment analysis module. The sentiment analysis employs an internal sentiment dictionary, where we selected the "Opinions (English)" sub-dictionary. The SPSS Modeler Text Analytics function offers researchers many complex and nuanced ways to study language. For this research, our aims were modest and we selected a relatively simple dichotomy to guide our analysis: all messages were coded as either positive, negative, or N/A by the Modeler software based on the appearance of certain sentimental words. Another commonly used sentiment dictionary known as AFINN, which was developed by Finn Årup Nielsen, has 2477 words and phrases, with each one rated on an ordinal scale of +5 to −5 (e.g. obliterate is −2, where rejoicing is +4) (Hansen et al. 2011; Ngoc and Yoo 2014).

Based on discussions with IBM officials, we feel comfortable that the Modeler dictionary likely bears some resemblance to the AFINN dictionary, though the exact parameters of the Modeler product are proprietary. For additional validation, we also ran the Twitter messages through the AFINN dictionary using the Urban Attitudes software, with its new sentiment analysis module.

Next, we sought to better explore the way in which the Twitter messages engaged with the topics of urban policy. While one would not expect Twitter users' posts to frequently relate to perceptions of place, we did seek to understand the extent to which this might be the case. We conducted a keyword text search for a set of 24 terms that pertain to New Bedford, their neighborhoods, or other topics related to child and family policy in a shrinking city.

Through a grounded approach, we developed the set of keywords for the qualitative analysis. The keywords were developed through a close reading of meeting minutes from six city boards and commissions. A research assistant reviewed the meeting minutes of the School Committee, Zoning Board, Planning Commission, City Council, Board of Health, and City Council from the years 2008–2014 and wrote a summary narrative for each year (n = 300). The researchers reviewed the narratives and the research team identified 24 frequently appearing terms relevant to urban and youth policy and planning.

## What We Found in Our Test

For six weeks from February 2014 to April 2014, we used the Urban Attitudes software to download 122,186 messages. Of those, 87,079 had valid fields and were tested for sentiment. The messages were collected in a relatively uninterrupted time basis into a series of individual *.csv files. We compiled those *.csv files into a single file and ran it through the SPSS Modeler Text Analytics program. The result was that 6268 (7.2 %) of the messages were classified as positive (including several variations of the concept of positive, including positive attitude, positive budget, positive competence, positive feeling, positive feeling emoticon, ☺, and positive functioning). A total of 4825 (5.5 %) were classified as negative (along with conceptually close variations of negative).

Those familiar with Twitter might not be surprised at how few messages evoked any sentiment: a total of 9 % (subsequent chapters will show that was a relatively low sentiment percentage). Mostly, people use Twitter to communicate about their favorite sports team, what they are having for dinner, or their plans for the evening. However, these results show that over 10,000 messages did have an embedded sentiment, and it was significantly more positive than negative.

To check these results, we also ran the Twitter messages through the custom-designed sentiment analyzer that used AFINN (also part of the "Urban Attitudes" software). Here the unit of analysis is the word, whereas the Modeler software used the tweet as the unit of analysis. The Urban Attitudes software searches for the appearance of positive and negative words preprogrammed into the AFINN dictionary; when a positive word appears, it becomes a marker for the presence of positive sentiment. By only using the AFINN dictionary and none of contextual information embedded in the Modeler software, the program found 58,490 positive words and 44,981 negative words (among a total word count of 1,139,761). Presented as a percentage, that means 5.1% positive words and 3.9% negative words, not too different than the Modeler results. Each AFINN word is weighted, so with the added weights (between –5 and +5), the positive words had a combined score of 132,838 and the negative words had a weighted combined score of –111,529.

Turning now to the keyword search, we did a Control-F search through all the Twitter messages to determine the number of appearances of our list of 24 keywords (see Table 3.1). Here we found that the words "schools," "zoning," "health," "safety," and "parks" appeared most frequently (see Table 3.2).

The work above constituted an initial step toward contributing to emergent literature around the use of social media data in planning by asking the following questions: (1) What are the strengths and weaknesses of microblogging data and (2) what is the range of uses for such data in planning practice and research? What we found was: *(1) Microbloggers generate a vast amount of data. (2) Microblog content parallels current events in the news, politics, and matters of local policy. (3) Microbloggers infrequently use sentiment and rarely microblog about these places. (4) The magnitude of microblogging data and the ability to capture it readily and a growing number of sophisticated analysis techniques may mean that quantity can make up for lack of quality. (5) Microblogging has the potential to join the set of imperfect descriptive measures of citizens' perspectives. Moreover, microblogging is unique among these data sets in that it may be used to go*

**Table 3.1** Keywords used in meeting minute and Twitter messages searches

| | |
|---|---|
| 1. | Children |
| 2. | School |
| 3. | Preschool |
| 4. | Safety |
| 5. | Vacant lot/vacancy |
| 6. | Housing violation |
| 7. | Field |
| 8. | Truant |
| 9. | Redevelop |
| 10. | Parks |
| 11. | Guns |
| 12. | Alcohol |
| 13. | Underage drinking |
| 14. | Violence |
| 15. | Foreclosure |
| 16. | Zoning |
| 17. | Health |
| 18. | Demolish |
| 19. | Smoke/smoking |
| 20. | Tobacco |
| 21. | Condemned |
| 22. | Lead |
| 23. | Prenatal |
| 24. | Health |

**Table 3.2** Appearance of keywords in meeting minutes and Twitter messages (top 10)

| | *Minutes* | | *Twitter* | |
|---|---|---|---|---|
| | # | *Rank* | # | *Rank* |
| School | 346 | 1 | 1602 | 1 |
| Zoning | 227 | 2 | – | – |
| Health | 140 | 3 | 247 | 3 |
| Safety | 121 | 4 | 33 | 9 |
| Parks | 75 | 5 | 47 | 7 |
| Housing violation | 67 | 6 | – | – |
| Smoke | 42 | 7 | – | – |
| Field | 39 | 8 | 305 | 2 |
| Children | 35 | 9 | 76 | 5 |
| Vacant | 25 | 10 | – | – |

*beyond describing citizens to helping develop an understanding of how people feel about the places they live and react to and even participate in public events going on around them.*

Although this analysis is very preliminary, it is also intriguing. We collected the Twitter feed for a short period of only four weeks. This does not allow for a truly robust analysis, especially given the ambitions of "Big Data" analysis. It is possible that any data set would contain the keywords at similar frequency and that this is simply a reflection of common English language usage. However, it could be that the pattern identified above reveals that, in both formal and informal settings, residents engage topics in similar ways. That would suggest that the informal world of communication is an extension of, and not necessarily an alternative to, formal communication, and vice versa. However, given the limitations of this analysis, this is entirely speculative. Reviewing the data across larger time frames and comparing it to other urban settings will help further test these ideas.

Additionally, an obvious question remains: How would one classify the majority of Twitter communications, since it appears that in the case of New Bedford in 2014, most tweets could not be classified as either positive or negative? A conventional content analysis might help illuminate how people use Twitter to communicate and how sentiment fits into this modern means of expression.

While this was only our first foray into urban social listening, it helped us to develop the necessary software tools, methodological frameworks, and analytical strategies that we use in the next two chapters. The limitations enumerated above did not completely go away, but for the next chapters we review two research projects that sought to build on the New Bedford research in a systematic and rigorous way—addressing the weaknesses illuminated in this chapter, though not quite eliminating them.

## References

Gaber, J., & Gaber, S. (2007). *Qualitative analysis for planning & policy: Beyond the numbers.* Chicago: APA Planners Press.

Hansen, L. K., Arvidsson, A., Nielsen, F. Å., Colleoni, E., & Etter, M. (2011). Good friends, bad news—Affect and virality in Twitter. The 2011 International Workshop on Social Computing, Network, and Services (SocialComNet).

Ngoc, P. T., & Yoo, M. (2014). The lexicon-based sentiment analysis for fan page ranking in Facebook. In *Information Networking (ICOIN), 2014 International Conference on*, IEEE, 444–448.

CHAPTER 4

# A Close Look at Urban Immigrant Communities

**Abstract** Immigrant communities have been the subject of much urban social science research, but this population can be difficult to identify and study effectively. Here we propose and evaluate a method for sentiment analysis of microblogging data of Portuguese-speaking residents. We ask, how do Portuguese-speaking microbloggers compare to English-speaking microbloggers in terms of overall happiness and attitude? And how do the results of a sentiment analysis of microblog data compare with traditional indicators of well-being in cities? We compare the results of the qualitative sentiment analysis with traditional quantitative indicators of economic and social well-being based on US Census data. While there are weaknesses in comparing overall sentiment between English and Portuguese, the benefits are vast in setting up statistically valid and reliable comparable frameworks.

**Keywords** New England • Portuguese • Linguistics • English

The previous chapter illustrates how microblogging data might improve the study of urban communities. Given New Bedford's status as a major immigrant hub, the results in the previous chapter raise a key question that warrants further investigation: how might the sentiment analysis of microblogging data function in multilingual immigrant communities?

© The Editor(s) (if applicable) and The Author(s) 2016
J.B. Hollander et al., *Urban Social Listening*,
DOI 10.1057/978-1-137-59491-4_4

Immigrant communities have been the subject of much urban social science research because they typically face a number of economic, social, and political challenges. In order to identify patterns and recommend and implement policy changes that are more inclusive of immigrants, it is necessary to be able to assess and measure the well-being of members of these communities. Currently, census data and economic indicators, as well as traditional surveys, interviews, and case studies, are the most common and effective ways to study and assess these communities. However, these methods have their limitations, including being slow and costly to implement, their difficulty in capturing a multilingual community, and the fact that, especially with many cities hosting undocumented immigrants, this population can be difficult to identify and study effectively.

The purpose of this chapter is to propose and evaluate a method for supplementing traditional indicators of well-being, such as demographic and economic data, with sentiment analysis of microblogging data. This chapter will specifically focus on Portuguese-speaking immigrant communities in Massachusetts Gateway Cities, using microblogging data collected from four of these cities and analyzed for sentiment in Portuguese and English.

"Gateway Cities" is a term used to identify cities that, during the Industrial Revolution and into the early twentieth century, served as major destinations for immigrants to the USA, often because they could offer skilled labor to the textile mills or other industries located there. The term encompasses predominantly older industrial cities that today have decaying infrastructure and limited economic resources with which to assist newcomers. In 2008 the chief executives of eleven cities signed a compact identifying these locations as "Gateway Cities," including: Fall River, New Bedford, Brockton, Fitchburg, Lawrence, Lowell, Haverhill, Springfield, Holyoke, and Pittsfield. The goal of this compact is to encourage these cities to work together to lessen the economic gap between them and rest of the state (UMass Dartmouth 2014a). In 2010, the Massachusetts legislature officially codified the term (Foreman and Larson 2014).

These post-industrial cities contend with a number of economic and social disadvantages. For example, 30% of Massachusetts residents living below the poverty line live in Gateway Cities, even though these cities account for only 15% of the total population of the state (UMass Dartmouth 2014b). According to census data for New Bedford, 19.2% of the population is foreign born, and 60.2% of foreign-born residents are limited English proficient (LEP). The median household income for

New Bedford is $36,789, while for Massachusetts as a whole the median household income is $66,658 (US Census Bureau 2014c). Several of the Gateway Cities are also among the cities in Massachusetts with the highest numbers of immigrants (Immigrant Learning center 2010). In fact, the foreign-born population of many Gateway Cities is trending back toward the percentage of the early twentieth century (Foreman & Larson).

As a case study of the immigrant population of these Gateway Cities, this research focuses on Portuguese-speaking residents. This is a diverse group, including people born in or with ancestry from Portugal, Cape Verde, and Brazil, along with smaller numbers from other Portuguese-speaking countries (Massachusetts Alliance of Portuguese Speakers 2008). Portuguese speakers as a group are especially relevant to focus on because Massachusetts has more Portuguese speakers than any other state, based on an American Community Survey Data, 2006–2010 estimate (Massachusetts Alliance of Portuguese Speakers 2012). The largest numbers of foreign-born Portuguese speakers in Massachusetts are originally from Brazil. In fact, a report from the Immigrant Learning Center found Brazil to be the top country of origin for all foreign-born residents of Massachusetts in 2009 (Immigrant Learning Center 2010). Portuguese-speaking immigrants also have significant historical ties to Massachusetts, as they have been immigrating to Massachusetts since the 1800s (Bailey 2000). While the country of origin has shifted—earlier immigrants were primarily from Portugal or the Azores—this group has had a long-standing cultural impact on certain areas of Massachusetts. There is significant overlap between Gateway Cities and Portuguese-speaking immigrant communities, with several of the Gateway Cities among the cities with the highest concentrations of Portuguese speakers, including Fall River, New Bedford, Lowell, and Brockton (Massachusetts Alliance of Portuguese Speakers 2008). These four cities are the focus of this study.

We seek to contribute to the study of the well-being of immigrant communities in Massachusetts by assessing Portuguese-speaking immigrants in Gateway Cities based on a sentiment analysis of microblogging data. In order to do so, we collect geotagged microblogging data using the Twitter Streaming API from the four heavily Portuguese-speaking Gateway Cities in Massachusetts. We also collected microblogging data from a larger region of Eastern Massachusetts and three control locations. We then conducted a sentiment analysis in English and Portuguese of these collected microblogs using sentiment expressed through social media as an indicator of well-being. Finally, we compared the results of the qualitative sentiment analysis with traditional quantitative indicators of economic and social well-being based on US Census data. Our research questions include:

1. How do Portuguese-speaking microbloggers in Massachusetts Gateway Cities compare to English-speaking microbloggers in terms of overall happiness and attitude?
2. How do the results of a sentiment analysis of microblog data compare with traditional indicators of well-being in Massachusetts Gateway Cities?

We focused on five collection areas: four Gateway Cities, including Brockton, New Bedford, Fall River, and Lowell, and a larger area of Eastern Massachusetts. We also collected data from three additional control locations for comparison purposes: Cambridge, Massachusetts; São Paulo, Brazil; and Lisbon, Portugal. Cambridge is home to both the Massachusetts Institute of Technology and Harvard University, universities which attract sizable student and faculty from around the globe, including Portuguese speakers. São Paulo and Lisbon are both cities where Portuguese is the primary language, each representing very different linguistic communities.

We collected geotagged tweets from all five collection areas for three periods in December 2014 and January 2015. Other studies based on the Twitter API collected data continuously for extended periods of time—9 weeks (Gordon 2013), 445 days (Lovelace et al. 2014), or 6 months (O'Connor et al. 2010). But unlike these studies, our primary purpose is simply to compare between locations and languages, not changes in sentiment over time. To this end, we consolidated all of our data sets from each time period into one file per location. See Table 4.1 for details.

The collected data is stored in a .CSV format. Each individual tweet has a unique ID number, a number assigned to the account holder, and a time and date stamp. tweets collected from updated versions of the program also contain geographic coordinates, a city name, and a column that indicates the language listed on the account of the user who published that particular tweet.

In order to direct the data collection, we put specific locations in the Urban Attitudes software using latitude and longitude. For Fall River and Brockton, we collected tweets from the smallest box we could get around

**Table 4.1** Data collection dates for each location

| *Location* | *Data collection dates* |
|---|---|
| New Bedford | December 1–20, January 6–19, January 20–30 |
| Brockton | December 1–20, January 6–19, January 20–30 |
| Fall River | December 1–20, January 6–19, January 20–30 |
| Lowell | January 21–February 16 |
| Eastern MA | December 1–20, January 6–19, January 20–30 |
| Harvard and MIT | January 26–February 16 |
| São Paulo | January 20–30 |
| Lisbon | January 20–30 |

each city boundary using a 0.1 degree (or 6-minute) grid. To collect data for Fall River, for example, we used the northeast bounding corner at the coordinates N 41 degrees, 48 minutes, (41.8) and W 71 degrees. The southwest bounding corner is at N 41 degrees and 36 minutes (41.6) and W 71 degrees and 12 minutes (71.2). As you can see from the attached map (Figs. 4.1 and 4.2), this includes small portions of several towns that surround Fall River. While collecting Twitter data in this way is imprecise, it will still help us approximate the attitudes of Fall River residents.

Once the stream of geotagged tweets was collected, we conducted a sentiment analysis using a sentiment analysis lexicon (or dictionary) called SentiStrength. This lexicon is available in both English and Portuguese, making it possible to compare sentiment between tweets in each language on the same scale. Furthermore, this lexicon is academically respected, having been evaluated in a number of peer-reviewed articles and used in multiple academic studies (Thelwall et al. 2011; Pfitzner et al. 2012; Zheludev et al. 2014; Kucuktunc et al. 2012; Weber et al. 2012; Garas et al. 2012; Grigore and Rosenkranz 2011; Vural et al. 2012; Giannopoulos et al. 2012).

The SentiStrength sentiment lexicon is simply a list of common words that are deemed to have sentiment, with a number value between −5 and +5 applied to each word; lower values reflect more negative emotions and higher values reflect more positive emotions. The Urban Attitudes program then searched the data set of collected tweets and matched words from the dictionary with words in the tweets. A total tally was reported,

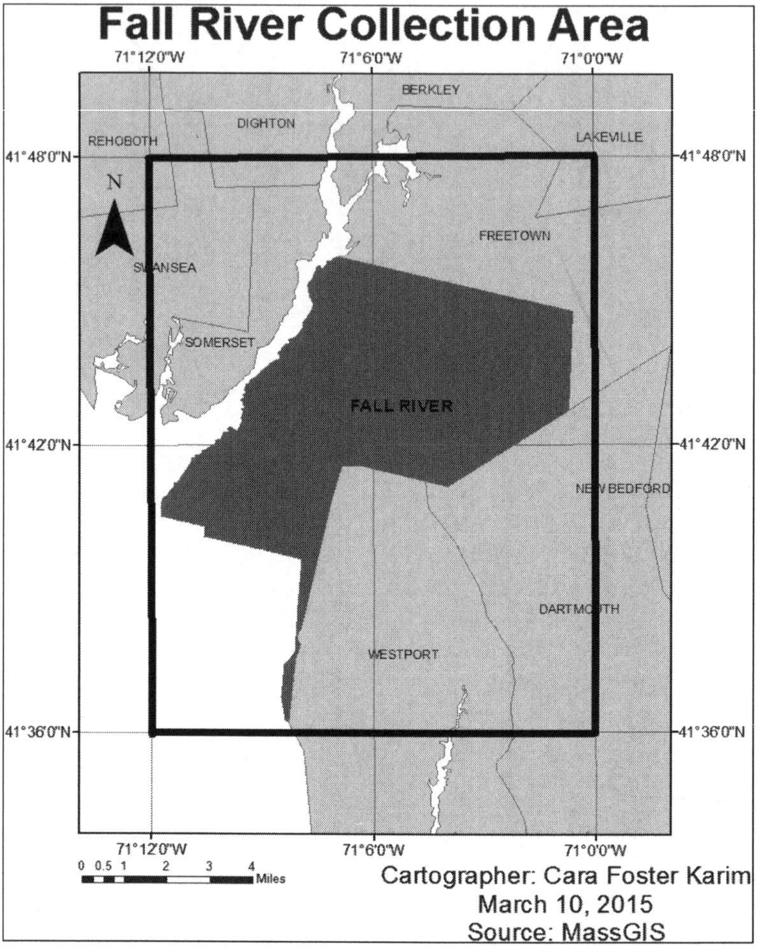


**Fig. 4.1** Fall River with latitude and longitude

with the number of times each positive or negative word appears, which was then multiplied by each word's sentiment score to produce an overall sentiment score for each data set. In order to compare English and Portuguese language sentiment on the same set of data, the program was run two different times, one with each dictionary, and the final scores were recorded.

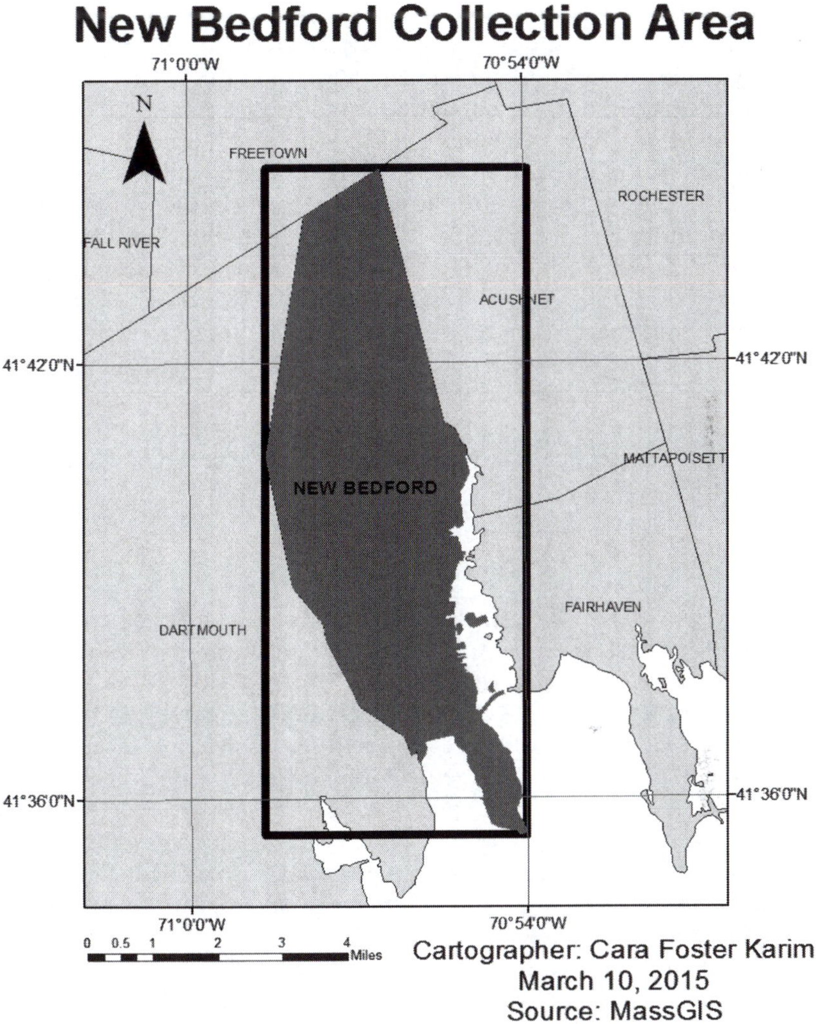


**Fig. 4.2** New Bedford with latitude and longitude

## Statistical Tests

We used a two-sample *T*-test to test whether the difference between the percentages of positive tweets was significant across different locations and between each language at each location. Then, in order to see if there was any correlation between the results of the sentiment analysis and demographic and socioeconomic variables, we created a ranking system for each city being studied and conducted a bivariate ordinal correlation test (Spearman's) to see how the variables related to each other.

As an addition check, we hand coded a sample of 300 of the collected and analyzed tweets in each language for the Brockton collection location and compared our assessment to the results of the sentiment analysis from the SentiStrength sentiment lexicon. We found that both dictionaries were correct in the overall sign of the sentiment (positive or negative) 75% of the time, although we often disagreed with the dictionaries' measurements of intensity for a given tweet. For the Portuguese dictionary, we found that 20% of the tweets flagged as having Portuguese sentiment were actually in Portuguese. Another 20% of the tweets were in Spanish, and the rest were in English or other languages. Due to the many cognates between Portuguese, Spanish, and English, it is impossible to eliminate all the erroneous tweets.

## Results and Analysis

In this section, we will first present the initial results from all of the data collection locations. Next we will discuss the statistical significance of differences in the results between different languages and between the data collection locations. Finally, we will present demographic information based on US Census data for the Massachusetts collection locations and discuss some correlations between the demographic data: the results of our analysis.

In total across all the data sets, we collected and analyzed 1,167,176 tweets. Some of our collection areas overlapped; so we calculated that we had, at minimum, 814,315 discrete tweets. Our primary results are presented in Tables 4.2 and 4.3.

The first major difference is that our coverage is significantly higher in the English than Portuguese, ranging from 52% to 58% across all the locations. This reflects both the larger number of words in the English sentiment dictionary and that Massachusetts is an area with an English-speaking majority. The second major thing to note is that the percentage of positive tweets is also much higher in each collection area.

The following table shows our results from the three control locations (Lisbon, São Paulo) and the Harvard and MIT area of Cambridge (Table 4.4).

**Table 4.2** Portuguese results for primary data collection areas

| *Location* | *Total tweets* | *Tweets w/ sentiment* | *Coverage (%)* | *% Positive* | *% Negative* | *Sentiment score* | *Normalized score* |
|---|---|---|---|---|---|---|---|
| Eastern MA | 315,448 | 169,086 | 53.6 | 56.5 | 43.4 | 61,200 | 0.3038 |
| Fall River | 30,770 | 17,895 | 58.2 | 54.9 | 45.2 | 1570 | 0.0730 |
| Brockton | 273,854 | 144,171 | 52.6 | 56.5 | 43.5 | 48,358 | 0.2826 |
| New Bedford | 37,732 | 21,783 | 57.7 | 53.9 | 46.1 | –1717 | –0.0649 |
| Lowell | 38,663 | 22,037 | 57.0 | 54.6 | 45.4 | 2561 | 0.0968 |

**Table 4.3** English results for primary data collection locations

| *Location* | *Total tweets* | *Tweets w/ sentiment* | *Coverage (%)* | *% Positive* | *% Negative* | *Sentiment score* | *Normalized score* |
|---|---|---|---|---|---|---|---|
| Eastern MA | 315,448 | 4277 | 1.3 | 50.8 | 49.2 | –959 | –0.2146 |
| Fall River | 30,770 | 217 | 0.7 | 50.7 | 49.3 | –26 | –0.1176 |
| Brockton | 273,854 | 4211 | 1.5 | 50.9 | 49.0 | –958 | –0.2184 |
| New Bedford | 37,732 | 249 | 0.6 | 50.8 | 49.2 | –48 | –0.1905 |
| Lowell | 38,663 | 350 | 0.9 | 50.8 | 49.2 | –89 | –0.2445 |

Note that in Lisbon and São Paulo, the coverage percentage in Portuguese is much higher, indicating a much larger pool of tweets in Portuguese. However, it is still not as high as the coverage in English in Massachusetts, probably due to limitations of the sentiment dictionary. Interestingly, the Cambridge collection area had the highest amount of positive sentiment in English but the lowest in Portuguese. Also, the Lisbon and São Paulo locations were more positive overall than the Massachusetts locations when measured in Portuguese, but less positive when measured in English. If measuring by the normalized score, however, the English tweets from São Paulo were more positive in sentiment than those from Cambridge. Also, the Lisbon and São Paulo locations had an even higher percentage of positive tweets overall than the Massachusetts locations when measured in Portuguese, but less positive when measured in English.

**Table 4.4** Results from control locations

| *Location* | *Language* | *Total tweets* | *Tweets w/ sentiment* | *Coverage (%)* | *% positive* | *% negative* | *Sentiment score* | *Normalized score* |
|---|---|---|---|---|---|---|---|---|
| Lisbon | Portuguese | 22,156 | 7774 | 35.1 | 55.0 | 45.0 | –452 | –0.0521 |
| São Paulo | Portuguese | 438,048 | 154,920 | 35.4 | 56.7 | 43.3 | 26,723 | 0.1514 |
| Harvard and MIT | Portuguese | 10,505 | 134 | 1.3 | 44.4 | 55.6 | –44 | –0.3259 |
| Lisbon | English | 22,156 | 2797 | 12.6 | 45.3 | 54.7 | 1297 | 0.4355 |
| São Paulo | English | 438,048 | 63,966 | 14.6 | 48.9 | 51.1 | 60,408 | 0.8937 |
| Harvard and MIT | English | 10,505 | 5533 | 52.7 | 62.2 | 37.8 | 5504 | 0.8508 |

## Significant Results

In order to determine which of these interesting differences might be statistically significant, we ran a series of two-sample T-tests. We considered differences in the same language between each location and then looked at differences between the results in each language at each location. For example, at 56.67% positive, the results from São Paulo were significantly more positive than the Portuguese results from every other location, with a p-value less than .05 for Fall River, New Bedford, and Lowell. The difference was even more significant, with a p-value less than .01 for Eastern MA, Brockton, Lisbon, and the Harvard and MIT location. Lisbon was also significantly ($p < .01$) different in positivity from all the New England locations except for Fall River, New Bedford, and Lowell. We found that, in both English and Portuguese, the results were significantly different between most of the Gateway Cities and most of the control locations. When comparing the Gateway Cities with each other, however, there were no statistically significant differences in Portuguese positivity. In the English results, there was no significant difference between Lowell and Fall River and New Bedford, although Brockton was significantly different than the others. With the exception of the Brockton English results, these findings validate our decision to group these four cities together for analysis purposes, since the differences between them in terms of sentiment scores are, for the most part, not statistically significant. See Appendices 2 and 3 for full tables of relationships between the results from all the data collection areas in each language.

We also tested within each location to see if there was a significant difference between the Portuguese and English positivity scores (see Table 4.5).

We found that the difference between the Portuguese and English percentages of positivity was statistically significant for the overall region of Eastern Massachusetts, Fall River, Lisbon, São Paulo, and Cambridge. In Eastern Massachusetts, Fall River, and Cambridge, the English results had a significantly higher percentage of positive tweets. In Lisbon and São Paulo, however, the percentage of positive tweets in English was significantly lower than those in Portuguese. We also aggregated the results for the four Gateway Cities and found the difference between their average scores to also be significant, with results in English significantly more positive than results in Portuguese.

Table 4.5 Significance of differences between languages in each location

| *Location* | *Portuguese positivity (%)* | *English positivity (%)* | *Z-score* | *p-value* |
|---|---|---|---|---|
| Eastern MA | 50.8 | 56.5 | 7.6716 | 0.0000[a] |
| Brockton | 50.7 | 54.9 | 1.2535 | 0.1056 |
| Fall River | 50.9 | 56.5 | 7.3034 | 0.0000[a] |
| New Bedford | 50.8 | 53.9 | 0.9963 | 0.1587 |
| Lowell | 50.8 | 54.6 | 1.4444 | 0.0749 |
| Lisbon | 55.0 | 45.3 | –9.1318 | 0.0000[a] |
| São Paulo | 56.7 | 48.9 | –34.6467 | 0.0000[a] |
| Harvard and MIT | 44.4 | 62.2 | 4.2122 | 0.0000[a] |
| **Average for all four Gateway Cities** | **50.8** | **55.0** | **–7.1297** | **0.0000[a]** |

[a]Indicates significance at $p < 0.05$

## Demographic Variables

To further explore the potential meaning in these results, we added demographic variables for each of the Massachusetts cities from the US Census American Community Survey 5-year estimates (2009–2013 and 2006–2010) (Table 4.6). These demographic variables are not intended to exactly represent the population of the data collection areas but rather provide some background and a rough approximation. The US Census boundaries for each of these cities are not the same as those we have used for our data collection areas. For the most part, our collection areas were slightly larger than the official city boundaries, including areas around each target city. For the Cambridge location, because we were trying to focus on the Harvard and MIT area, only about two-thirds of the area of the city is included in our collection area, although this two-thirds includes the most densely populated area of the city.

These selected variables are often used as indicators of socioeconomic status or well-being, especially median household income or percent of adult population with a high school diploma. We chose percent foreign born and language other than English spoken at home as a way to measure the immigrant communities in these cities. Unfortunately, data on Portuguese speakers is not available at the city level, but there is American Community Survey data on people who reported Portuguese or Brazilian ancestry. Brazilian or Portuguese ancestry is an imperfect approxima-

**Table 4.6** Key demographic variables

| *Demographic variable* | *Brockton* | *Fall River* | *New Bedford* | *Lowell* | *Average of Gateway Cities* | *Cambridge* | *Massachusetts overall* |
|---|---|---|---|---|---|---|---|
| Population | 94,089 | 88,697 | 95,078 | 108,861 | 96,681 | 107,289 | 6,745,408 |
| % under 18 | 25.7 | 21.5 | 23.20 | 23.7 | 23.7 | 11.4 | 20.8 |
| % 65 and over | 11.9 | 15.1 | 14.6 | 10.1 | 13.0 | 9.5 | 14.8 |
| % Black/African American | 31.2 | 3.9 | 6.4 | 6.8 | 12.0 | 11.7 | 8.1 |
| % Hispanic | 10.0 | 7.4 | 16.7 | 17.3 | 12.7 | 7.6 | 10.5 |
| % foreign born | 24.7 | 19.0 | 19.9 | 24.8 | 22.2 | 27.7 | 15.0 |
| Median household income | $49,025 | $33,211 | $35,999 | $49,452 | $41,922 | $72,529 | $66,866 |
| % with high school diploma | 80.5 | 70.3 | 70.5 | 78.8 | 75.0 | 93.7 | 89.4 |
| Language other than English spoken at home (%) | 36.6 | 34.0 | 37.0 | 41.9 | 37.4 | 32.1 | 21.9 |
| Population density | 4398.40 | 2681.90 | 4754.30 | 7842.10 | 4919.00 | 16,470.20 | 839.4 |
| Portuguese ancestry | 3946 | 41,443 | 38,603 | 5308 | 22,325 | 2208 | 311,767 |
| Brazilian ancestry | 1238 | 1197 | 997 | 2901 | 1583 | (No data) | 65,170 |
| % of population with Brazilian or Portuguese ancestry | 5.5 | 48.1 | 41.6 | 7.5 | 24.7 | 2.0 | 5.6 |

Sources: Decennial Census and American Community Survey 5-year estimates, 2006–2010 and 2009–2013 (US Census Bureau 2014a, b, 2015a, b, c, d, e)

tion for Portuguese speakers, as many residents may be of Portuguese or Brazilian heritage but not currently speak Portuguese. These data also ignore any residents of Cape Verdean background, as they are not tracked as a specific ancestry group by the census. Nonetheless, Brazilian and Portuguese ancestry is the best proxy for Portuguese speakers available in the census data.

The four Gateway Cities we studied have a lower median household income, a younger population, fewer high school diplomas, and a higher percentage of foreign-born residents than observed for Massachusetts as a whole. Furthermore, 24.7% of Gateway City residents claimed Brazilian or Portuguese ancestry—far higher than the statewide average of 5.6%. See Appendix 4 for a table of some of these key variables for the segment of the population that indicated Brazilian or Portuguese ancestry in the four Gateway Cities.

Although representing almost a quarter of the population of these Gateway Cities, there are some key differences between residents of Portuguese or Brazilian background and the rest of the population. The number of residents of Portuguese or Brazilian ancestry who are under 18 is higher than average, while the number who are 65 or over is much lower. This group is also almost twice as likely to be foreign born. They are somewhat less likely to have a high school diploma, although this differed between those of Portuguese and Brazilian heritage—Brazilians had a higher than average high school graduation rate, and Portuguese had a lower than average high school graduation rate.

## Demographic Correlations

In order to assess any correlations between our results and the demographic data, we first ranked each of the five Massachusetts collection areas on select demographic variables, as well as how they performed in the results of the sentiment analysis. Then, a Spearman's rho test was run on these ranked cities to look at correlations between the sentiment variables and the demographic variables. We use the percent of positive and negative tweets in Portuguese and English, as the overall score for each language, and the normalized score for each language as our sentiment variables. For demographic variables, we chose population, percent foreign born, percent with a high school diploma or higher, median income, percent under 18 and over 65, population density, and Portuguese and Brazilian ancestry. The rankings are displayed in Tables 4.7 and 4.8.

**Table 4.7** Massachusetts city rankings based on demographic variables

| *City* | *Population* | *Foreign born* | *Median income* | *High school education* | *Under 18* | *65 and over* | *Population density* | *Portuguese ancestry* | *Brazilian ancestry* |
|---|---|---|---|---|---|---|---|---|---|
| Brockton | 4 | 3 | 3 | 2 | 1 | 3 | 4 | 4 | 2 |
| Fall River | 5 | 5 | 5 | 5 | 4 | 1 | 5 | 2 | 3 |
| New Bedford | 3 | 4 | 4 | 4 | 3 | 2 | 3 | 1 | 4 |
| Lowell | 1 | 2 | 2 | 3 | 2 | 4 | 2 | 3 | 1 |
| Cambridge | 2 | 1 | 1 | 1 | 5 | 5 | 1 | 5 | 5 |

**Table 4.8** Massachusetts city rankings based on results of sentiment analysis

| *City* | *Percent positive* | *Percent negative* | *Raw sentiment score* | | *Normalized score* | | | |
|---|---|---|---|---|---|---|---|---|
| Language | English | Portuguese | English | Portuguese | English | Portuguese | English | Portuguese |
| Brockton | 2 | 1 | 4 | 5 | 1 | 5 | 2 | 3 |
| Fall River | 3 | 3 | 3 | 2 | 4 | 1 | 4 | 1 |
| New Bedford | 5 | 4 | 1 | 3 | 3 | 3 | 5 | 2 |
| Lowell | 4 | 2 | 2 | 4 | 2 | 4 | 3 | 4 |
| Cambridge | 1 | 5 | 5 | 1 | 5 | 2 | 1 | 5 |

We found statistically significant ($p < .05$) correlations between some of the demographic and sentiment variables. Having a high population of residents who were 18 years of age and under correlated positively with positive Portuguese sentiment scores, whereas a high population of residents with Portuguese ancestry correlated negatively with positive English sentiment scores. Conversely, a high population of residents with Brazilian ancestry correlated positively with positive Portuguese sentiment scores. A high percentage of residents who had at least completed high school correlated positively with a higher normalized sentiment score in English and negatively with a higher normalized sentiment score in Portuguese.

## Conclusions

Our urban social listening methods allowed us to discover some potentially important patterns, trends, and relationships. Our study is not without its limitations (which we explore in greater depth in Chap. 6), but it does offer several new and potent insights that are only possible through this kind of qualitative examination of microblogging data.

We chose to study just two linguistic communities: English speaking and Portuguese speaking. However, in Massachusetts Gateway Cities, dozens of languages are spoken by sizable portions of the population, including Spanish, French, Chinese, Japanese, Vietnamese, and Creole. Without much more effort, urban researchers can bring in SentiStrength (or other dictionaries) for each language. We discovered many weaknesses in comparing overall sentiment between English and Portuguese, though with the assistance of professional linguists, these barriers could be overcome.

Building on the research described in this chapter, we next sought to look more broadly across space and time. In the following chapter, we present research which includes both Twitter and a more established, survey-based indicator of well-being, the American Housing Survey, comparing and contrasting between different time periods and different US cities.

## References

Bailey, B. (2000). History and description of Portuguese immigration and the East Providence/SE New England Portuguese community. University of Massachusetts—Amherst. http://works.bepress.com/benjamin_bailey/63. Accessed 31 Mar 2015.

Foreman, B., & Larson, S. (2014). Going for growth: Promoting immigrant entrepreneurship in Massachusetts Gateway Cities. MassINC Gateway Cities Innovation Institute. http://www.massinc.org/~/media/Files/Mass%20Inc/Research/Full%20Report%20PDF%20files/GoingforGrowth_PromotingImmigrantEntrepreneurshipinMassachusettsGatewayCities.ashx. Accessed 1 Dec 2014.

Garas, A., Garcia, D., Skowron, M., & Schweitzer, F. (2012). Emotional persistence in online chatting communities. *Scientific Reports*, 2, article 402.

Giannopoulos, G., Weber, I., Jaimes, A., & Sellis, T. (2012). Diversifying user comments on news articles, web information systems engineering (WISE 2012). *Lecture Notes in Computer Science 2012*, 100–113.

Gordon, J. (2013). *Comparative geospatial analysis of Twitter sentiment data during the 2008 and 2012 US Presidential elections.* Eugene: University of Oregon.

Grigore, M., & Rosenkranz, C. (2011). Increasing the willingness to collaborate online: An analysis of sentiment-driven interactions in peer content production. ICIS 2011 Proceedings. Paper 20.

Immigrant Learning Center (2010). Massachusetts immigrant entrepreneurs: Engines for economic growth, wealth and job creation. The Immigrant Learning Center, Inc. http://www.ilctr.org/wp-content/uploads/2010/11/ILC-Profiles-web-2.pdf. Accessed 1 Dec 2014.

Kucuktunc, O., Cambazoglu, B. B., Weber, I., & Ferhatosmanoglu, H. (2012). A large-scale sentiment analysis for Yahoo! Answers, In *Proceedings of the 5th ACM International Conference on Web Search and Data Mining.*

Lovelace, R., Malleson, N., Harland, K., & Birkin, M. (2014). *Geotagged tweets to inform a spatial interaction model: A case study of museums.* Leeds: School of Geography, University of Leeds.

Massachusetts Alliance of Portuguese Speakers (2008). In Massachusetts. http://www.maps-inc.org/our-community/portuguese-speaking-community-in-massachusetts/. Accessed 6 Nov 2014.

Massachusetts Alliance of Portuguese Speakers (2012). Community health profile. http://www.maps-inc.org/wp-contenuploads/2012/05/Community-Health-Profile.pdf. Accessed 6 Nov 2014.

O'Connor, B., Balasubramanyan, R., Routledge, B. R., & Smith, N. A. (2010). From tweets to polls: Linking text sentiment to public opinion time series. *ICWSM, 11*, 122–129.

Pfitzner, R., Garas, A., & Schweitzer, F. (2012). Emotional divergence influences information spreading in Twitter. *ICWSM, 12*, 2–5.

Thelwall, M., Buckley, K., & Paltoglou, G. (2011). Sentiment in Twitter events. *Journal of the American Society for Information Science and Technology, 62*(2), 406–418.

UMass Dartmouth (2014a). The urban initiative at UMass Dartmouth. http://www.umassd.edu/media/umassdartmouth/seppce/centerforpolicyanalysis/urbaninitiative/assets/urbaninitiative.pdf. Accessed 6 Nov 2014.

UMass Dartmouth (2014b). Plan for renewal. http://www.umassd.edu/media/umassdartmouth/seppce/centerforpolicyanalysis/urbaninitiative/assets/planforrenewal.pdf. Accessed 6 Nov 2014.

United States Census Bureau (2014a). Fall River (city) State and County QuickFacts. http://quickfacts.census.gov/qfd/states/25/2523000.html. Accessed 3 Dec 2014.

United States Census Bureau (2014b). New Bedford (city) State and County QuickFacts. http://quickfacts.census.gov/qfd/states/25/2545000.html. Accessed 6 Nov 2014.

United States Census Bureau (2014c). Selected characteristics of the native and foreign born populations, 2008–2012. American Community Survey 5-year estimate.

United States Census Bureau (2015a). Cambridge (city) State and County QuickFacts. http://quickfacts.census.gov/qfd/states/25/2511000.html. Accessed 11 Mar 2015.

United States Census Bureau (2015b). Lowell (city) State and County QuickFacts. http://quickfacts.census.gov/qfd/states/25/2537000.html. Accessed 11 Mar 2015.

United States Census Bureau (2015c). Brockton (city) State and County QuickFacts. http://quickfacts.census.gov/qfd/states/25/2509000.html. Accessed 11 Mar 2015.

United States Census Bureau (2015d). Table B01003. 2006–2010 American Community Survey Selected Population Tables.

United States Census Bureau (2015e). 2009–2013 American Community Survey Selected Population Tables.

Vural, G., Cambazoglu, B. B., & Senkul, P. (2012). Sentiment-focused web crawling. In *Proceedings of the 21st ACM international conference on information and knowledge management*, 2020–2024.

Weber, I., Ukkonen, A., & Gionis, A. (2012). Answers, not links: extracting tips from Yahoo! answers to address how-to web queries. In *Proceedings of the fifth ACM international conference on Web search and data mining* (WSDM '12).

Zheludev, I., Smith, R., & Aste, T. (2014). When can social media lead financial markets? *Scientific Reports*, *4*. doi:10.1038/srep04213.

CHAPTER 5

# A National Comparison: Twitter versus the American Housing Survey

**Abstract** We examine differences in resident perceptions of neighborhood quality of life, as well as expressed positive and negative sentiment while accounting for changes in population among cities between 1970 and 2010. We find no evidence that population loss leads to a lower evaluation of life satisfaction. Additionally, we find while tweets are a source for consistently determining the positive and negative affect of individuals on a geographic basis and that people generally have a positive feeling about their neighborhood, there are no significant relationships between the Twitter data sets and traditional ones. Thus, planners or policy makers should not presume that a singular measure will provide a complete picture of well-being.

**Keywords** American Housing Survey • Quality of life • Neighborhood attitudes • Population decline • Shrinking cities

Declining cities in North America and Europe have increasingly attracted attention over the last decade, as municipal leaders, planners, and researchers have grappled with the question of how to confront the issue of declining population. At its root, this represents a conflict between the traditional conception of urban growth as a primary barometer of success with the stark reality that many cities may simply never reach their previous peak population. Despite popular perception, including sensa-

© The Editor(s) (if applicable) and The Author(s) 2016

J.B. Hollander et al., *Urban Social Listening*,
DOI 10.1057/978-1-137-59491-4_5

tionalist stories in the media about cities like Detroit that have certainly seen better days, initial research suggests that the link between negative population growth and a lower quality of life is not so clear-cut as some might expect (Hollander 2011; Delken 2008). Accordingly, it should not be taken for granted that residents of cities prefer population growth over stability or even decline (Van Dalen and Henkens 2011).

Armed with an awareness that the power of urban social listening is becoming increasingly apparent, we sought to examine the subjective well-being (SWB) experienced by residents in a broad selection of US cities, testing the general assumption that it is relatively unpleasant to live in cities that are not growing as compared with those that are. In particular, we investigated eight cities nationwide, determining the degree to which there is a correlation between having flat or declining population and having a lower level of reported or measured SWB. These cities represent a cross section of cities in the USA with regard to population changes over the last half century. We accept that the sample size is less than ideal, but it is a starting point upon which future work may productively build. Moreover, we were limited by available data. These specific cities were chosen because they were covered in a database of several million tweets gathered by Dr. Jeff Nichols of IBM starting in 2013, as well as by the US Census Bureau's American Housing Survey covering the years from 2007 to 2011.

This chapter examines two primary research questions:

1. Are there differences in resident perceptions of neighborhood quality of life, as well as expressed positive and negative sentiment, when considering population change among cities between 1970 and 2010?
2. Is Twitter a reliable gauge of resident perceptions when compared with existing measures of well-being, namely, resident satisfaction with their neighborhoods?

We use two primary data sources to help shed light on these questions. The first is a series of tweets. These were collected separately over two multiweek periods; the first beginning in the fall of 2013, and the second during the spring of 2015. We then imported these tweets into the Urban Attitudes software program to measure differences in the level of sentiment for residents of different cities.

The second data set comes from the American Housing Survey (AHS). Of particular relevance to this study, the AHS asks respondents to rate their neighborhood on a scale of 1–10, which we aggregate to compute an average sentiment score for each city in our sample.

In answer to question #2, we find little relationship between SWB as measured by Twitter posts and average neighborhood satisfaction as measured by the AHS. However, this does not mean that social media is inherently flawed as a vehicle for measuring sentiment. Instead, it highlights that social well-being is a multidimensional and complex phenomenon. Our sentiment analysis of text is measuring an aspect of SWB, while questions about neighborhood satisfaction, or other aspects of life satisfaction, speak to an entirely different aspect of SWB. We conclude that both traditional surveys and sentiment analysis have valuable, separate roles to play in assisting with the determination of SWB across different populations.

## Methods

We use a lexicon-based sentiment analysis of tweets powered by the AFINN sentiment dictionary. The AFINN dictionary was developed by Finn Årup Nielsen and has been used in multiple research studies: including the identification of anti-vaccine sentiments from tweets (Brooks 2014), evaluation of more than 5000 advertisements in business magazines (Abrahams et al. 2013), as part of a model predicting fluctuations in global currency markets (Jin et al. 2013), and, not to mention, the research presented in Chap. 3 of this book. AFINN ranks words on an ordinal scale ranging from +5 to −5. For example, "abusive" is given a score of −3, while "satisfied" is given a score of +2. The latest version of AFINN has 2477 words and is capable of capturing variants of words, such as recognizing "loooooove" as "love." We use AFINN to score all sentiment-containing words and develop an overall average for each study city.

We acquired tweets over two separate multiweek periods, allowing us to somewhat account for the undue influence of specific news events on public sentiment. The first period is between November 26, 2013, and January 20, 2014, using a database provided to us by Dr. Jack Finn of IBM. The second includes tweets collected between March 3 and 19, 2015, using our own Urban Analysis software package. For simplicity's sake, we labeled the first batch of tweets, including data from January 2014, as "2013 tweets" and the second batch as "2015 tweets." While 2015 tweets is a much smaller data set, it nonetheless contains a sufficient number of tweets for comparison, as demonstrated by the remarkable consistency between the two data sets. Figure 5.1 shows a small set of sample tweets as displayed within the UA software, exported to Excel in .csv file format. Column A contains an anonymous identifying number for the user instead of his or her actual username.

| ◇ | A | B | C | D | E |
|---|---|---|---|---|---|
| 1 | 162965776 | I'm so tired | -84.388977 | 33.756307 | 2015-03-03 15:32:51 |
| 2 | 230573079 | Aye man I need to go get my piercing today fr ! | -84.441818 | 33.681245 | 2015-03-03 15:32:53 |
| 3 | 468611502 | GOOD MORNING | -84.369766 | 33.745663 | 2015-03-03 15:33:06 |
| 4 | 13729242 | Microsoft Will Offer a Peek at SharePoint 2016 at Ignite by @druadh20 | -84.345716 | 33.734866 | 2015-03-03 15:33:08 |
| 5 | 154316034 | @_ImBoldBitchhh They had more people. I didn't go  but I heard | -84.412031 | 33.750896 | 2015-03-03 15:33:09 |
| 6 | 81472795 | @_naptural_ @TheyCallMeEli__ Lmaoo ive never heard anything like it | -84.509793 | 33.843409 | 2015-03-03 15:33:17 |
| 7 | 468611502 | My birthday tomorrow | -84.376104 | 33.744145 | 2015-03-03 15:34:33 |
| 8 | 2247244138 | Why tf are we learning science in world history | -84.372097 | 33.780193 | 2015-03-03 15:34:47 |
| 9 | 235235379 | 21 st birthday I will be bringing the city out | -84.403843 | 33.708725 | 2015-03-03 15:35:21 |
| 10 | 139604119 | Mmmmmm I do have a type | -84.403984 | 33.707372 | 2015-03-03 15:35:21 |
| 11 | 756673417 | Niggas always make promises but break em | -84.489138 | 33.698938 | 2015-03-03 15:36:42 |
| 12 | 43463937 | What what you mean you can't?  Whay is wrong with these people | -84.442184 | 33.765766 | 2015-03-03 15:36:49 |

**Fig. 5.1** Sample tweets from Atlanta

We used the geographic information systems software platform ARCGIS coupled with digit municipal boundary files from the US. Census Bureau to determine a minimum bounding rectangle for each city. The UA program then pulls out all geotagged tweets with latitude and longitude coordinates falling within each bounding rectangle.

The minimum bounding rectangle is only an approximation for municipal boundaries. There will always be some tweets captured from outside the formal city boundaries. Figure 5.2 demonstrates this, show-

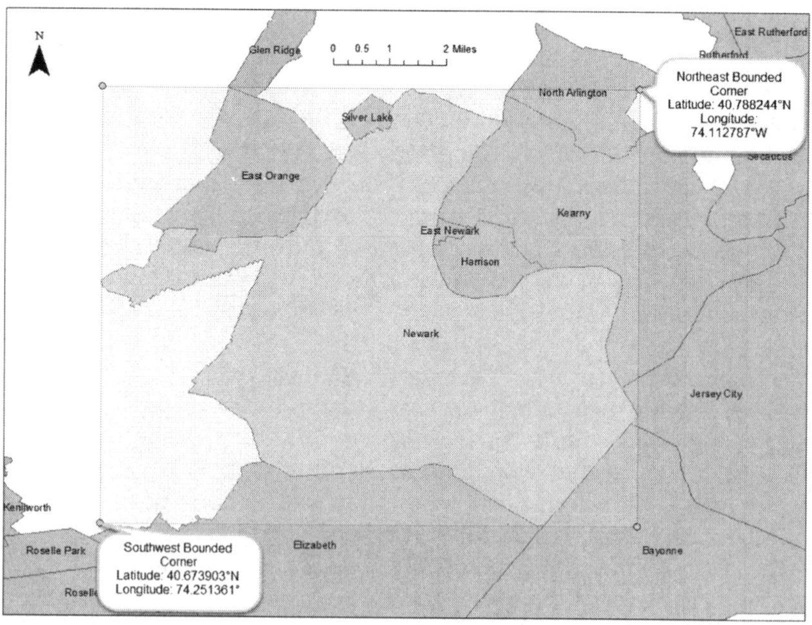


**Fig. 5.2** Newark city boundaries and coordinates

ing the bounding rectangle for the City of Newark, NJ, against its civic boundaries.

Our other primary data set is the AHS. The US Census Bureau conducts the AHS every two years for selected metropolitan areas. It also distinguishes respondents living in the central city from others living in the metropolitan area. Among other things, the AHS asks respondents to rate their neighborhood on a 1–10 scale. We take the numeric average for all central city residents' overall rating of their neighborhood.

## Analysis and Results

Table 5.1 lists our eight study cities along with their respective population change between 1970 and 2010. Only two of the eight added population since 1970, Houston and Indianapolis. Providence was virtually unchanged, while Washington, DC, and Atlanta lost modest numbers. However, Newark, New Orleans, and St. Louis all lost a substantial share, with Newark losing nearly a third and St. Louis and New Orleans almost half of the 1970 population. New Orleans is a case of special concern, given the mass exodus following Hurricane Katrina in 2008. It is worth noting that New Orleans had already experienced considerable population loss even before Katrina, with roughly 140,000 fewer residents in 2005 than in 1970.

We calculate several statistics for each city, including the city's overall sentiment score, the average score per tweet. We also compute the total number of tweets, sentiment tweets, and percentage of positive tweets (Tables 5.2 and 5.3). Similar to Chap. 4, we measure average sentiment as the city's overall sentiment score divided by the total number of tweets for

**Table 5.1** Population change in the eight study cities

| *City* | *Population, 1970* | *Population, 2010* | *Change* | *Percent change (%)* |
|---|---|---|---|---|
| Atlanta | 496,973 | 420,003 | -76,970 | -15 |
| Houston | 1,232,802 | 2,099,451 | 867,461 | 70 |
| Indianapolis | 744,624 | 820,445 | 75,821 | 10 |
| New Orleans | 593,471 | 343,829 | -249,642 | -42 |
| Newark | 382,417 | 277,140 | -104,790 | -28 |
| Providence | 179,213 | 178,042 | -1171 | -1 |
| St. Louis | 622,236 | 319,294 | -302,942 | -49 |
| Washington, DC | 756,510 | 601,723 | -154,787 | -20 |

**Table 5.2** 2013 tweet sentiment analysis results

| *City* | *Overall score* | *Avg. score per tweet* | *Sentiment-containing tweets* | *Total tweets* | *% positive sentiment tweets* |
|---|---|---|---|---|---|
| Atlanta | 971,693 | 0.2108 | 2,526,884 | 4,608,671 | 37.66 |
| Houston | 1,168,144 | 0.2005 | 3,101,918 | 5,827,597 | 35.96 |
| Indianapolis | 567,103 | 0.3142 | 1,035,167 | 1,804,805 | 40.11 |
| New Orleans | 243,410 | 0.0977 | 1,336,069 | 2,492,313 | 35.30 |
| Newark | 680,821 | 0.2060 | 1,735,069 | 3,305,276 | 35.59 |
| Providence | 105,352 | 0.1893 | 309,133 | 556,671 | 37.46 |
| St. Louis | 318,451 | 0.2130 | 842,460 | 1,494,820 | 38.55 |
| Washington, DC | 435,869 | 0.1596 | 1,462,535 | 2,731,275 | 36.00 |

that city. We calculate the "percent positive sentiment tweets" as the share of total tweets collected from each city that contain at least one word with positive sentiment. Appendix 5 includes additional data details.

All cities had a positive overall score in 2013 (Table 5.2), and only New Orleans had a negative score in 2015 (Table 5.3). Indianapolis was the most "positive" city in both time periods according to both measures. The second most positive was St. Louis. The least positive city was New Orleans. However, even in the case of New Orleans in 2015, there were more positive-sentiment-containing tweets (57,260) than negative-sentiment-containing tweets (48,903), but the negative words were more extreme on the AFINN polarity spectrum. There were no cases where a city had more negative sentiment tweets than positive.

The choice of whether to use average sentiment score per tweet or the percentage positive sentiment tweets turns out to be largely irrelevant to the overall results. When measured on an ordinal scale, both metrics produce a near identical rank ordering of cities. The average sentiment score is also highly correlated with percentage positive score in both years, with correlation coefficients of 0.82 for 2013 and 0.94 for 2015.

There are significant differences in our summary sentiment scores over time. The citywide averages of both measures declined from 2013 to 2015: from .199 to .117 for average sentiment and from 37% to 34% for percent positive tweets. This is not altogether unexpected or problematic. Other studies show that sentiment ratings from Twitter data are sensitive to current events (Balduini et al. 2013). Nor is this necessarily evidence that the general mood is on a declining trend. We would need far more

**Table 5.3** 2015 tweet sentiment analysis results

| *City* | *Overall score* | *Avg. score per tweet* | *Sentiment-containing tweets* | *Total tweets* | *% positive sentiment tweets* |
|---|---|---|---|---|---|
| Atlanta | 18,702 | 0.1255 | 76,819 | 149,010 | 35.19 |
| Houston | 28,572 | 0.0469 | 313,556 | 609,335 | 33.60 |
| Indianapolis | 27,150 | 0.2355 | 63,268 | 115,311 | 37.60 |
| New Orleans | −13,514 | −0.0738 | 89,875 | 183,195 | 31.26 |
| Newark | 4126 | 0.0687 | 31,007 | 60,035 | 34.00 |
| Providence | 6015 | 0.1807 | 17,132 | 33,288 | 34.71 |
| St. Louis | 8912 | 0.2281 | 20,356 | 39,078 | 35.96 |
| Washington, DC | 29,181 | 0.1228 | 120,351 | 237,626 | 33.99 |

**Table 5.4** Paired difference of means T-tests, 2013 and 2015

| | *Average sentiment score* | *Percent positive tweets* |
|---|---|---|
| 2013 | 0.199 | 0.371 |
| 2015 | 0.117 | 0.345 |
| Difference | −0.082 | −0.025 |
| Degrees of freedom | 7 | 7 |
| t stat | 3.378 | 10.134 |
| P-value (two tailed) | 0.012 | 0.000 |
| Pearson's R (across years) | 0.766 | 0.924 |

data measured at different time periods to test whether this is the case. Despite the overall decline in sentiment, the degree of decline remains relatively consistent across the different cities. The average sentiment score has a Pearson's R correlation coefficient of .77 between 2013 and 2015. The consistency of the percent positive tweet measure is even stronger with a correlation coefficient of .924 between 2013 and 2015 (see Table 5.4). The relative ranking of our sample cities is also relatively consistent, with Spearman's Rho statistics of .74 (average sentiment) and .93 (percent positive). We see this consistency as supporting evidence of the reliability of tweets as a source for determining the positive and negative affect of individuals on a geographic basis.

Next, we turn to sentiment as measured according to the AHS. The AHS asked respondent to rank their neighborhood on a scale of 1–10. We tallied these scores to produce a citywide average score (Table 5.5). Because the AHS surveys different cities in different years, Table 5.5 reports each city's score for the most recent year of data available.

Overall, residents are pretty happy with their neighborhoods. On a scale of 1–10, the average scores all rank in the low-to-mid 7s—suggesting that they are far more pleased with their neighborhood than not. Furthermore, all eight cities fall within a rather narrow band of just +/−0.77 points.

For our final set of comparisons, we test whether there are observable differences in satisfaction measured by the AHS versus those based on a sentiment analysis of mined Twitter data. We also consider a number of additional variables collected from the US Census's Quick Fact files, based on their American Community Survey (see Appendix 5 for details). While by no means a comprehensive list, these variables are commonly associated with happiness and well-being: such as wealth, age, and education. We convert each into ordinal rankings (1 through 8), to facilitate comparison across such vastly different units of measure (Table 5.6), and use the Spearman's Rho correlation coefficient to measure the degree of association.

We find no significant relationships between AHS neighborhood satisfaction and tweet-based sentiment at standard p-value confidence thresholds of 90 and 95 % (Table 5.7). However, this may be largely the consequence of such a small sample size leading to the conclusion of a false negative. There are modest positive ordinal correlations between AHS neighborhood satisfaction and the average tweet score in 2015 (Rho = .50) as well

**Table 5.5** AHS survey results

| *City* | *State* | *Survey year* | *Neighborhood opinion* |
|---|---|---|---|
| Atlanta | GA | 2011 | 7.337 |
| Houston | TX | 2007 | 7.376 |
| Indianapolis | IN | 2011 | 7.357 |
| New Orleans | LA | 2011 | 7.707 |
| Newark | NJ | 2009 | 7.000 |
| Providence | RI | 2011 | 7.522 |
| St. Louis | MO | 2011 | 7.400 |
| Washington | DC | 2007 | 7.768 |

*Source*: US Census Bureau

**Table 5.6** Ordinal rankings of AHS, tweet and demographic data

| | *Atlanta* | *Houston* | *Indianapolis* | *New Orleans* | *Newark* | *Providence* | *St. Louis* | *Washington, DC* |
|---|---|---|---|---|---|---|---|---|
| AHS opinion | 3 | 6 | 1 | 5 | 8 | 2 | 7 | 4 |
| Average tweet score, 2013 | 3 | 5 | 1 | 8 | 4 | 6 | 2 | 7 |
| Average tweet score, 2015 | 4 | 7 | 1 | 8 | 6 | 3 | 2 | 5 |
| Percent positive tweets, 2013 | 3 | 6 | 1 | 8 | 7 | 4 | 2 | 5 |
| Percent positive tweets, 2015 | 3 | 7 | 1 | 8 | 5 | 4 | 2 | 6 |
| Median income | 2 | 3 | 4 | 6 | 8 | 5 | 7 | 1 |
| Percent under 18 years | 7 | 1 | 3 | 5 | 2 | 4 | 6 | 8 |
| Percent bachelor's degree or higher | 2 | 5 | 7 | 3 | 8 | 6 | 4 | 1 |
| Population change, 1970 to 2010 | 4 | 1 | 2 | 7 | 5 | 3 | 8 | 6 |
| Percent foreign born | 6 | 2 | 5 | 8 | 3 | 1 | 7 | 4 |
| Mean travel time to work | 4 | 3 | 7 | 6 | 1 | 8 | 5 | 2 |
| Median value of owned homes | 3 | 6 | 8 | 5 | 2 | 4 | 7 | 1 |
| Poverty rate | 5 | 6 | 7 | 4 | 1 | 2 | 3 | 8 |

**Table 5.7** Ordinal correlations (Spearman's Rho), AHS neighborhood scores versus Twitter sentiment

| | *AHS neighborhood satisfaction* | |
|---|---|---|
| *Twitter sentiment metric* | *Rho* | *Pr* > \|*r*\| |
| Average tweet score, 2013 | 0.0952 | 0.823 |
| Average tweet score, 2015 | 0.5000 | 0.207 |
| Percent positive tweets, 2013 | 0.5238 | 0.183 |
| Percent positive tweets, 2015 | 0.3810 | 0.352 |

as with the percent positive tweets in 2010 (.52, respectively)—but not for the other two Twitter-based metrics. Furthermore, we do see some consistent patterns between the AHS and Twitter-based sentiment. Indianapolis was the highest-ranking city both for the AHS and for all four tweet-based sentiment measures. Atlanta and Houston also had generally consistent rankings. But there are also counterexamples. Namely, St. Louis, which was the second lowest-ranking city in AHS neighborhood satisfaction but was the second highest in tweet-based sentiment. The results for Providence and Newark are also somewhat mixed. In sum, our analysis suggests a possible association between the two metrics, but we cannot rule out this association is purely the result of sampling variation. More data collected over more cities is necessary.

Table 5.8 shows a similar correlation analysis between our sentiment metrics and a number of socioeconomic indicators. By and large, the AHS measure of neighborhood opinion and our Twitter sentiment scores were unrelated to the socioeconomic condition of cities. Although the low sample size puts us at serious risk of finding false negatives. Ignoring statistical significance. Only one relation was statistically significant—the strong negative association between neighborhood satisfaction and commute length—and then only at a 90% confidence level. Most people hate long commutes, so it makes perfect sense that travel distance to work is a key source of neighborhood dissatisfaction. The Twitter-based measures of sentiment show no statistically significant association with commute length, although several do have negative correlations in excess of .4. There is also a modestly high positive correlation between neighborhood opinion and median household income, although falling just shy of the 90 percent two-tailed confidence threshold. Again, income is not associated with our tweet-based sentiment measures. Median home value

**Table 5.8** Ordinal correlations (Spearman's Rho), sentiment scores versus socioeconomic factors

| | *AHS neighborhood opinion* | | *Average tweet score, 2013* | | *Average tweet score, 2015* | | *Percent positive tweets, 2013* | | *Percent positive tweets, 2015* | |
|---|---|---|---|---|---|---|---|---|---|---|
| | *Rho* | *Pr > \|r\|* | *Rho* | *Pr > \|r\|* | *Rho* | *Pr > \|r\|* | *Rho* | *Pr > \|r\|* | *Rho* | *Pr > \|r\|* |
| Population change, 1970–2010 | 0.45 | 0.26 | 0.21 | 0.61 | 0.10 | 0.82 | 0.17 | 0.69 | 0.10 | 0.82 |
| Median income | 0.55 | 0.16 | -0.14 | 0.74 | 0.02 | 0.96 | 0.21 | 0.61 | -0.07 | 0.87 |
| Percent under 18 years | -0.21 | 0.61 | 0.17 | 0.69 | -0.19 | 0.65 | -0.24 | 0.57 | -0.12 | 0.78 |
| % bachelor's degree or higher | 0.10 | 0.82 | -0.40 | 0.32 | -0.21 | 0.61 | -0.05 | 0.91 | -0.29 | 0.49 |
| Percent foreign born | 0.10 | 0.82 | -0.12 | 0.78 | 0.00 | 1.00 | -0.10 | 0.82 | -0.10 | 0.82 |
| Mean travel time to work | -0.67 | 0.07 | -0.10 | 0.82 | -0.48 | 0.23 | -0.43 | 0.29 | -0.36 | 0.39 |
| Median value of owned homes | -0.17 | 0.69 | -0.52 | 0.18 | -0.40 | 0.32 | -0.48 | 0.23 | -0.40 | 0.32 |
| Poverty rate | -0.45 | 0.26 | 0.00 | 1.00 | -0.07 | 0.87 | -0.26 | 0.53 | 0.02 | 0.96 |

is the only socioeconomic condition that appears to be associated with tweet-based sentiment, and then only for a single year (2013). Although below conventional statistical thresholds, there are modest negative correlations between median home value and average tweets and percent positive tweets in 2013. It is hard to explain why people in cities with more high-valued homes might be increasingly prone to dissatisfaction, as home values are usually associated with wealth and thus greater life satisfaction. It could be that with many homes losing value in the recession, residents felt additional stress, which began to abate by 2015. Again, this requires further study.

Most importantly for this book, we found no significant relationship between resident satisfaction and long-term population change (1970 and 2010). We do find a somewhat modest positive correlation (.45) between neighborhood satisfaction and how a city ranks according to past growth, but there is also a 26% chance this association is erroneous. Even so, dissatisfaction with one's neighborhood is not necessarily the same as dissatisfaction with one's life. The correlations between population loss and tweet-based sentiment scores, which are presumably more indicative of overall subjective life satisfaction, are very weak. It could be that residents dislike the physical remnants of decline in their neighborhoods, such as abandoned buildings and boarded-up windows, but are still generally satisfied.

## Discussion

This project has demonstrated that, for eight major cities in the USA, there exists no significant correlation between population gain or loss in that city between 1970 and 2010 and the subjective well-being of the residents of that city, as measured by the AHS and millions of total tweets. For our first research question, "Are there differences in resident perceptions of neighborhood quality of life, as well as expressed positive and negative sentiment, when accounting for changes in population among cities between 1970 and 2010?" the answer is "probably not," though with such a small sample size and the results of a 0.45 positive correlation, we certainly have not proven anything here instead have shown that potential power of microblogging data to shed light on an otherwise complex relationship. In answering our second research question, "In computing differences among cities, is Twitter a reliable gauge of resident perceptions when compared with more traditional measures of well-being?" we have come to the conclusion that the answer is "Yes," but we cannot point to significant statistical evidence beyond the literature to substantiate this statement. There was no statistically significant correlation between AHS results and tweet results, at least as far as ordinal ranking of cities is concerned, though we did find very similar rankings for three of our sample cities: Indianapolis, Atlanta, and Houston. In this study, we were limited by the availability of historic city-level tweets matching those in the AHS sample. Future studies should collect data on a much larger cohort of municipalities to determine test for meaningful differences.

Moreover, psychological literature suggests that the AHS and tweets were measuring two different aspects of subjective well-being, life satisfaction and affect. AHS asks residents to rate their neighborhood from 1 to 10, and this rating is reflective of only a piece of how a respondent might answer the larger question of how he or she would rate his or her overall life satisfaction. tweets, in contrast, measure the positive and negative affect that Twitter users demonstrate via the content of their posts. Although the relevant literature certainly indicates that there is a relationship between life satisfaction and affect, they would not occupy the exact same space in a Venn diagram of SWB, so the lack of a statistically significant relationship between AHS and tweets is not proof that Twitter is without merit in indicating at least one aspect of SWB.

Future research might consider making use of multiple dictionaries for sentiment analysis, as well as incorporating other traditional measures of happiness. An in-depth study of one city, with manual scoring of a body of tweets, might provide additional insights. Should time and funding permit, short interviews with residents of cities from which tweets are being gathered could directly ask about overall life satisfaction rather than relying on AHS data that present an incomplete picture of this metric.

To that end, future research using microblogging data as a corpus for sentiment analysis should strongly engage with the psychological literature in considering what kind of happiness or SWB is being measured. A promising line of inquiry may involve interviews with positive psychologists to outline different forms of SWB in greater depth, with an objective of determining how surveys, polls, text analysis, and other bodies of data shine a light on particular forms. The first wave of sentiment analysis research has been completed, and now is the time to develop a more rigorous understanding of what exactly sentiment analyses are telling us.

Finally, the ultimate purpose of performing these analyses must be mentioned. A review of urban planning literature that includes microblogging data indicates that the use of this data for planners is only at a very early stage. Sentiment analyses can inform public officials and employees about how residents are moving about their city, what they think of particular places in their city, and what service gaps exist for different populations. This research intimates that shrinking cities specifically are not places doomed to negativity and pessimism. Indeed, the cities among our cohort that have lost large chunks of their population base over the last several decades demonstrate that quality of life can be retained and enhanced, and there may be a serendipitous relationship to be developed with positive psychologists who are capable not only of describing what different metrics of SWB are truly measuring but how those metrics might be developed over the long term.

## Limitations and Future Research

Simply put, no sentiment analysis will be perfect, and they necessarily involve subjective judgments. Thus, while sentiment analysis is a valid form of qualitative inquiry, it can always be improved, and this section will describe the limitations inherent in this chapter. Before moving on to the limitations that come from the difficulty of analyzing language, it must be acknowledged that Twitter is not necessarily a representative sample of the population at large and likely skews toward a younger user base (Gayo-Avello 2011). Therefore, certain demographic groups may be underrepresented, and results of a sentiment analysis conducted on tweets should be considered as one source of information augmenting other sources, such as traditional polling.

Among the language-based challenges is that of negation. For example, a tweet could describe something as "never good," which the Urban Attitudes software, using the AFINN dictionary, would treat as a tweet containing a positive sentiment ("good") with a value of +3. Clearly the intent of the writer, however, is to express a negative sentiment. There are many strategies employed to overcome the limitation of negation, of which Taboada et al. (2011) provide an excellent overview. One strategy, partially employed by the AFINN dictionary, is recognizing the two words "don't like" as a negative sentiment. However, the dictionary does not yet contain a sufficient number of these terms. Moreover, there are sentences in which the negation is more than one word away from the modifier, which requires algorithms that determine the polarity of an entire sentence rather than simply looking for sentiment-containing words.

A similar limitation is the use of rhetorical devices, a la, "I was expecting to love it," which would be classified very positively based on the word "love" but is actually expressing something much less positive if not outright negative (Mullen and Malouf 2006). The dictionary is also generally incapable of recognizing sarcasm, does not analyze emoticons, and cannot hope to cover the full range of slang employed in tweets, though it does include some slang words and a wide variety of vulgarities. A potential line of future inquiry involves generating manual counts of uses of sarcasm and other rhetorical devices and determining the degree to which the sentiment score computed by the dictionary is inaccurate. It may turn out that such instances do not significantly impact the results of a sentiment analysis.

The sentiment analysis software itself is admittedly imperfect, and its kinks are still being worked out. During the course of this research, as well as other work being done at the UAL, it was noticed that the software has a problem with contractions, considering the word "won't," for example, to be the word "won" followed by the letter "t." While this does not cause any difference as regard to the letter t, the word "won" is considered a positive sentiment and thus incorrectly contributes to the positive total for that city. Given the size of these tweet files—in some cases containing over a million tweets for a single city—it is infeasible to manually make corrections to contraction problems and is an issue for future efforts to work out.

An additional limitation, most clearly demonstrated in the case of Houston, with its multiple annexations of land along highway spurs leading into and out of the city, is that tweets are not restricted entirely to the central city in question. Some tweets in this analysis are certainly posted from outside the central city, although it is worth considering why this might or might not be important. Are tweets posted from half a mile outside of Atlanta or St. Louis unrepresentative of the affect experienced by residents of those cities? Such tweets may, in fact, come from residents of those cities. How would a study be devised to exclude tweets from non-residents, including only tweets from residents of the central city? For instance, there is no guarantee that a tweet from downtown Atlanta belongs to a resident of Atlanta as opposed to a tourist.[1]

To address this latter problem, future researchers might include only those tweets coming from users who consistently post from the city being investigated or, despite the potential unreliability of the location users volunteer on their Twitter profiles, by matching reported location with the latitude/longitude coordinates of tweets. Also worth consideration is that even tweets from tourists in New Orleans might fairly represent the affect that is, for whatever reasons, created or generated by New Orleans for people that experience life in one way or another in that city. Researchers should give pause when formulating research questions and give serious thought to what it is they are trying to measure or analyze.

For example, are we looking *only* into the affect experienced by residents of cities, regardless of the implications that shrinkage or growth might have on commuters, shoppers, and tourists? What implications do shrinkage and growth have in terms of creating and sustaining affect or life satisfaction that in turn attract or repel non-residents? How might we conceptualize cities in the context of their metropolitan statistical areas (MSAs) as determined by the US Census Bureau? Are there differences among growing or shrinking central cities when bearing in mind the growth or shrinkage of the suburbs and exurbs? Atlanta, a city that has technically shrunk since 1970 but that has seen incredible growth at the metro level, is a prime example. While these questions are beyond the work of this research, they are important considerations for future study.

## Implications for Urban Planning and Policy

With respect to our first research question, our findings confirm previous work suggesting that the relationship between decreasing population and SWB is not nearly as clear-cut as would be expected given popular perceptions. As a result, shrinking cities should not be considered places whose residents are doomed to declining quality of life, making those cities unworthy of investment. Indeed, a look back at the twentieth century indicates undeniably that many cities once thought to be beyond hope are quite capable of finding stability and resurgence. Strategies employed in cities that have already experienced a turnaround should be given a fair effort in shrinking cities, and, in any case, stability should be seen as an objective worthy of pursuit in its own right. The work of the positive psychologists holds significant promise here for urban planners and policy makers, and their research should be strongly regarded as a bridge between the fields of psychology and planning.

As to the second research question, the implications of Twitter for the future of urban planning and policy are nuanced but certainly affirm its value. It is not sufficient for planners or policy makers to presume that a sentiment analysis, or for that matter a more traditional measure of SWB such as a poll or survey, will provide a complete picture of the happiness of residents or visitors. The specific form of SWB, and the ways in which different forms will interact with one another and with the larger culture, must be considered at the outset. Though sentiment analysis can now be done relatively cheaply and quickly, great care should be taken to avoid simplistic or reductionist approaches and assumptions.

This is particularly relevant when one realizes that the different ways of measuring SWB will provide complementary information rather than operating as substitutes for one another. Planners and policy makers concerned with SWB should determine in advance what they are trying to measure in light of the ultimate aims of how their metrics will be used and ensure that the data they are gathering will truly speak to what they are attempting to measure. This task will be made easier in the future, as research into Big Data continues to generate improved means of conducting sentiment analysis, new ways of overcoming limitations, and the ability to eliminate noise and hone in on the specific sentiments that are of particular relevance to the research, planning, and policy-making endeavors under consideration. As this process unfolds, the value of tweets and other social media data sets for planners and policy makers stands to increase considerably.

## Note

1. Conversely, are only tweets from residents posted while those residents are in the city itself relevant to a sentiment analysis?

## References

Abrahams, A. S., Coupey, E., Zhong, E. X., Barkhi, R., & Manasantivong, P. S. (2013). Audience targeting by B2B advertisement classification: A neural network approach. *Expert Systems with Applications, 40*(8), 2777–2791.

Balduini, M., Della Valle, E., Dell'Aglio, D., Tsytsarau, M., Palpanas, T., & Confalonieri, C. (2013). Social listening of city scale events using the streaming linked data framework. In *The semantic web–ISWC 2013* (pp. 1–16). Berlin/Heidelberg: Springer.

Brooks, B. (2014). Using Twitter data to identify geographic clustering of anti-vaccination sentiments. University of Washington. https://digital.lib.washington.edu/researchworks/bitstream/handle/1773/26187/Brooks_washington_0250O_13777.pdf

Delken, E. (2008). Happiness in shrinking cities: A research note. *Journal of Happiness Studies, 9*(2), 213–218.

Gayo-Avello, D. (2011). Limits of electoral predictions using Twitter. In *Proceedings of the Fifth International AAAI Conference on Weblogs and Social Media*, AAAI, 490–493.

Hollander, J. (2011). *Sunburnt cities: The great recession, depopulation and urban planning in the American Sunbelt.* London/New York: Routledge.

Jin, F., Self, N., Saraf, P., Butler, P., Wang, W., & Ramakrishnan, N. (2013, August 11–14). Forex-foreteller: Currency trend modeling using news articles. In *Proceedings of the 19th ACM SIGKDD International Conference on Knowledge Discovery and Data Mining*, Chicago, 1470–1473.

Mullen, T., & Malouf, R. (2006). A preliminary investigation into sentiment analysis of informal political discourse. In *Proceedings of the AAAI-2006 Spring Symposium on Computational Approaches to Analyzing Weblogs*, 159–162.

Taboada, M., Brooke, J., Tofiloski, M., Voll, K., & Stede, M. (2011). Lexicon-based methods for sentiment analysis. *Computational Linguistics, 37*(2), 267–307.

Van Dalen, H. P., & Henkens, K. (2011). Who fears and who welcomes population decline? *Demographic Research, 25*(13), 437–464.

CHAPTER 6

# Conclusion

**Abstract** In this concluding chapter of the book, we discuss limitations to urban social listening in general and our research in particular, cover recommendations for improvement in future studies, synthesize the major themes that emerged, sketch out a methodological guide for researchers, and provide final thoughts. The chapter offers hope for using microblog data to overcome perennial limitations of other urban social science methods yet invites caution about the kinds of technical and conceptual problems that arise in doing this type of research.

**Keywords** Sentiment analysis • Twitter • Microblogs

For the three empirical studies presented in this book, we crafted research questions that we sought to answer using urban social listening, with an eye to providing a practical guide for conducting research using large social media data sets. While the answers began quite inconclusively in the New Bedford study (Chap. 3), they increased in clarity in the Portuguese research (Chap. 4) and gained further clarity in the multicity work presented in Chap. 5. As with any research, each of these studies had much in the way of limitations, but because of the new and novel data sources and analytical methods presented, each project provided promising points of entry for considering the potential and pitfalls of urban social listening.

© The Editor(s) (if applicable) and The Author(s) 2016

J.B. Hollander et al., *Urban Social Listening*,
DOI 10.1057/978-1-137-59491-4_6

This concluding chapter has a full agenda: we will first discuss limitations to urban social listening in general and our research in particular, cover recommendations for improvement in future studies, synthesize the major themes that emerged, sketch out a methodological guide for researchers, and provide final thoughts.

## Limitations

The three fundamental limitations for these studies were: the shortcomings of using Twitter itself as a research tool, the mistakes and inconsistencies within our research methods, and flaws we discovered in the sentiment lexicons that we used for the analysis. However, these limitations are worth examining individually especially with the understanding that research limitations are to be expected with relatively new methodologies, and only through deeper investigation will improvements be developed.

First, it is important to be aware of the inherent limitations of using microblog data for research purposes. Microbloggers are by no means a representative sample of the population of a geographic area and likely skew toward a younger user base (Gayo-Avello 2011). The tiny fraction of all tweets that are public, geotagged, and available from the Twitter Streaming API is even more limited. Furthermore, microblog data is by nature extremely poor in quality. Many of the tweets collected for this study are promotional ones from companies or organizations, or multiple retweets of the same content. It is also important to note that, at best, only about half of the tweets collected had any sentiment at all, as measured by our lexicon vocabulary.

There is also no easy way to tell, based on the data we collected, whether the Twitter user who posted a given message is actually a resident of the study area or simply passing through. It would be possible to look up the user ID of each user to see what location they had indicated as their hometown, but not everyone lists this information, and people frequently move and fail to update their location. This limitation was most clearly demonstrated in the case of Houston, with its multiple annexations of land along highway spurs leading into and out of the city, that tweets are not restricted entirely to the central city in question. Some tweets in this analysis are certainly posted from outside the central city, although it is worth considering why this might or might not be important. Are tweets posted from half a mile outside of Atlanta or St. Louis unrepresentative of the affect experienced by residents of those cities? Such tweets may, in fact, come from residents of those cities. How would a study be devised to exclude tweets from non-residents, including only tweets from residents of the central city? For instance, there is no guarantee that a tweet from downtown Atlanta belongs to a resident of Atlanta as opposed to a tourist. And what is the difference in the sentiments expressed by residents or visitors? Are they affected by place differently?

There were also numerous limitations to our data collection and analysis methods. We encountered some technical difficulties with the Urban Attitudes program that meant the data collection was often interrupted and was not constant.[1] Also, due to the change in information we had about the scale at which Twitter Streaming API data was available, some of the collection areas were much larger, including more portions of other cities and towns than others. Even for the areas where we did use the same method to delineate a collection area, geographic boundaries of cities varied in size. Also, due in part to technical issues and partly to adaptations of the research plans, we did not collect data from all locations for the same period or length of time. With regard to the research found in Chap. 5, a primary limitation was the low number of cities from which tweets were gathered. This limitation was caused by the inadequate availability of tweets from prior years[2] and meant that a more robust sample size of cities could not be produced.

Finally, there were a number of limitations that arose from the sentiment lexicons we used to analyze the collected data. We used multiple sentiment lexicons, including one in Portuguese and others in English. The reason for using the SentiStrength lexicon was that it was one of the only ones we were able to find that used the same scoring system in each language and had been evaluated for comprehensiveness and accuracy by academic sources. However, once we started to use it for the research in Chap. 4, we found that there were major differences and inconsistencies between the English and Portuguese lexicons, making it very hard to reliably compare results between them. First, the Portuguese dictionary contained far fewer words than the English one, and a much larger percentage of those words were scored as negative. This would potentially color the results by reducing the percentage of tweets that were found to have sentiment in Portuguese and making it more likely to identify more negative tweets. Second, the Portuguese lexicon included many English words, particularly slang words or abbreviations common in Internet usage such as "lol."

While it is likely true that Portuguese-speaking Twitter users commonly mix such English words into their Portuguese tweets, when applying the Portuguese sentiment lexicon to a predominantly English data set, a very high percentage of extraneous tweets were included in the analysis even though they were not actually in Portuguese at all. With both of the lexicons, we found through our qualitative content analysis that many common words were scored in a way did not reflect their most common usage. For example, in the English lexicon we found that the word "like" was given a fairly positive score, even though the vast majority of the time it appeared in the results, the actual meaning involved either making a comparison or simply using the word as filler, so that the overall content of the tweet was often not positive at all. This latter point likely extends to content analysis of English-only tweets.

Another significant language-based challenge is that posed by negation. For example, a tweet could describe something as "never good," which the Urban Attitudes software, using the AFINN dictionary, would treat as a tweet containing a positive sentiment ("good") with a value of +3. Clearly, the intent of the writer, however, is to express a negative sentiment. There are many strategies employed to overcome the limitation of negation, of which Taboada et al. (2011) provide an excellent overview. One strategy, partially employed by the AFINN dictionary, is recognizing the two words "don't like" as a negative sentiment. However, the dictionary does not yet contain a sufficient number of these terms. Moreover, there are sentences in which the negation is more than one word away from the modifier, which requires algorithms that determine the polarity of an entire sentence rather than simply looking for sentiment-containing words.

A similar limitation is the use of rhetorical devices, like, "I was expecting to love it," which would be classified very positively based on the word "love" but is actually expressing something much less positive if not outright negative (Mullen and Malouf 2006). The dictionary is also generally incapable of recognizing sarcasm, does not analyze emoticons, and cannot hope to cover the full range of slang employed in tweets, though it does include some slang words and a wide variety of vulgarities. A potential line of future inquiry involves generating manual counts of uses of sarcasm and other rhetorical devices and determining the degree to which the sentiment score computed by the dictionary is inaccurate. It may turn out that such instances do not significantly impact the results of a sentiment analysis.

The sentiment analysis software itself is admittedly imperfect, and its kinks are still being worked out. During the course of this research, as well as other work being done at the Tufts Urban Attitudes Lab, it was noticed that the software has a problem with contractions, considering the word "won't," for example, to be the word "won" followed by the letter "t." While this does not cause any difference as regard to the letter t, the word "won" is considered a positive sentiment and thus incorrectly contributes to the positive total for that city. Given the size of these tweet files—in some cases containing over a million tweets for a single city—it is infeasible to manually make corrections to contraction problems.

## Recommendations for Further Research

Research on digital self-expression, the sentiments it conveys, and the relationship to place has great potential. There are a number of ways in which future researchers could improve upon our methods in order to produce a study with fewer limitations. Beginning with the Portuguese-English study, comparing sentiment results between two different languages is inherently complicated due to different linguistic structures and cultural differences, but there are some steps that could be taken to mitigate these issues. First, using a computer program to identify the language of each tweet and sorting the results into two separate data sets before conducting any analysis would significantly eliminate the error associated with cognates in the sentiment lexicon, resulting in tweets being counted for Portuguese sentiment when not actually in Portuguese. Another solution for a researcher with a background in linguistics and psychology, and a longer time frame, would be to edit or rewrite both sentiment lexicons to ensure that they were as equivalent as possible in terms of numbers of words and the proportions of positive and negative words. This would require more technical computer programming knowledge.

An additional means of increasing the relevance and validity of the Chap. 4 study would be to conduct a traditional survey of Portuguese- and English-speaking residents in these Gateway Cities to supplement the results from the sentiment analysis and the census data. It would likely be beneficial to conduct a study in which Twitter sentiment results in different languages, in a series of different international and domestic contexts, would be compared, in an attempt to see how cultural differences or immigrant or native-born status impact overall well-being. Finally, increasing the sample size by including more cities and collecting data over a longer period of time would also potentially increase the validity of the results. Our research in Chap. 5 particularly would have benefited from the inclusion of a much larger number of cities.

To address the problem of acquiring tweets from residents of specific cities, future researchers might include only those tweets coming from users who consistently post from the city being investigated or, despite the potential unreliability of the location users volunteer on their Twitter profiles, by matching reported location with the latitude/longitude coordinates of tweets. Also worth consideration is that even tweets from tourists in New Orleans might fairly represent the affect that is, for whatever reasons, created or generated by New Orleans for people that experience life in one way or another in that city. Researchers should give pause when formulating research questions and give serious thought to what it is they are trying to measure or analyze.

For example, are we looking *only* into the affect experienced by residents of cities, regardless of the implications that shrinkage or growth might have on commuters, shoppers, and tourists? What implications do shrinkage and growth have in terms of creating and sustaining affect or life satisfaction that in turn attract or repel non-residents? How might we conceptualize cities in the context of their metropolitan statistical areas (MSAs) as determined by the US Census Bureau? Are there differences between growing and shrinking central cities when bearing in mind the growth or shrinkage of the suburbs and exurbs? Atlanta, a city that has technically shrunk since 1970 but that has seen incredible growth at the metro level, is a prime example. While these questions are beyond the work of this research, they are important considerations for future study.

## Synthesis

Across our three projects, several key themes arose. Of greatest importance is that tweets can be considered a complement to existing, imperfect measures of attitudes and opinions, such as surveys or interviews. Our research demonstrated similarities to other measures of sentiment, including public meeting minutes and census responses, while incorporating a far larger number of participants in an unobtrusive manner. This includes reaching communities that may not be present at public meetings or that may have additional challenges such as being undocumented immigrants or facing language barriers.

Criticism of social media data sets as being unrepresentative of the public at large is certainly valid, but less acknowledged is the ability of urban social listening to capture sentiments expressed by individuals who are typically left out by existing methods. Census surveys may be statistically valid, but capture only a tiny sliver of resident opinions of place, and are widely certainly used in public policy formulation. Public forums or meetings, on the other hand, draw attendance only from those most engaged in the political process, with turnout far smaller than even the already subpar voter turnout in municipal elections. There is much to be improved about urban social listening, but the promise of gaining near instantaneous insights and input from a large swath of otherwise excluded or difficult to access individuals should not be discounted.

Moreover, this methodology enables voices to be heard in an anonymous fashion, and those who wish to remain unreached are free to keep their Twitter feeds private (and to choose not to have their latitude and longitude coordinates published as part of their tweets), to say nothing of avoiding social media entirely. The cost of this form of research on a per-person basis is dramatically lower than surveys or interviews and is one of the greatest advantages of the urban social listening methodology. In short, large numbers of users may be included in a general scan of a geographic area, with the information about those users being as much or as little as they prefer, and with expenses kept to a minimum. In addition, the length of time required for data collection can be as little as a week or two and can be done from the lab, classroom, or at home, without the need for extensive travel or mailing.

## Methodological Guide

While certainly imperfect, the research experiences we have outlined in this book provide the basis for a defensible, clear, and useful setup steps a researcher can take in approaching the study of microblogging data.

Step 1, Articulate Research Questions: As with any other study, key research questions will guide the precise decisions around methods and data around urban social listening, but typically, an urban social listening research question will have a spatial element.

Step 2, Delineate Area of Interest: The Urban Attitudes software, along with other similar software programs, collects data from Twitter using a rectangle defined by the latitude and longitude of the northeast and southwest corners of any geography on Earth. While the researcher's specific area of interest may be smaller than that rectangle, at this stage the data must be collected at the larger geography (later, using a geographic information system, it is possible to filter out only tweets in that smaller geography).

Step 3, Set up tweet Acquisition: Using software like Urban Attitudes, begin to download tweets onto either a desktop or server. Depending on the nature of the study, special arrangements may need to be made to accommodate large volumes of data (it does not take long for a study to generate hundreds of gigabytes of Twitter data). While not discussed earlier in the book, there are a number of private online services that will sell researchers historic Twitter data. For well-financed projects, this option may be useful to consider.

Step 4, Analysis: The research reported here relied on keyword and sentiment analysis, grounded in larger content analysis techniques. Some projects may seek other analytical strategies, like tracking the movement of some users across time and space or linking overall sentiment to other data sources. A key part of any sentiment analysis is the selection (and refinement) of a lexicon or dictionary, choices need to be made in selecting the appropriate language(s) employed, as well.

Step 5, Validation: This step is the chance for the researcher to validate their findings by looking at other sources but also to attempt to improve on the insights that other sources provide. Validation can come through secondary data analysis of related sources (like US Census data) or through qualitative methods like surveys, interviews, or focus groups.

## Conclusion

That we were able to exhibit some consistency between the Twitter data and other data sources—despite the enumerated limitations—is rather remarkable. As these methods are perfected, planners, academics, and policy makers will have voluminous data to work with, and new light may be shed on broader public perceptions and attitudes about important issues in any given geographic area. Moreover, there will likely be new discussions about the citizens being reached by any given methodology, traditional or emerging—a welcome development for all who seek to understand public sentiments.

Urban social listening also provides an opportunity to search for significant relationships between the attitudes and opinions expressed by residents of different cities with demographic data collected via other means. Never before have the views of so many people, expressed about so many issues, been incorporated into statistical analyses of cities. Though our studies found few statistically significant correlations, this is a highly promising avenue of study and will only prove more valuable as city sample sizes grow in future work.

## Notes

1. For the research presented here, we ran the Urban Attitudes software from desktop computers—but in more recent projects, we have run the software from servers and have not had the same problems with interruptions.
2. Historical Twitter data has become increasingly available at coarse geographies and at high costs.

## References

Gayo-Avello, D. (2011). Limits of electoral predictions using Twitter. In *Proceedings of the Fifth International AAAI Conference on Weblogs and Social Media,* AAAI, 490–493.

Mullen, T., & Malouf, R. (2006). A preliminary investigation into sentiment analysis of informal political discourse. In *Proceedings of the AAAI-2006 Spring Symposium on Computational Approaches to Analyzing Weblogs*, 159–162.

Taboada, M., Brooke, J., Tofiloski, M., Voll, K., & Stede, M. (2011). Lexicon-based methods for sentiment analysis. *Computational Linguistics, 37*(2), 267–307.

# Appendix 1: Urban Attitudes Software

Urban Attitudes is a data mining and text analysis tool Justin Hollander and Dibyendu Das developed in 2013. The software supports the following operations:

- Analyzing a large variety of text files; and
- Downloading tweets from Twitter filtered by locations.

## Downloading Tweets

The program requires a set of tokens from Twitter to download tweets. This can easily be obtained by signing up on Twitter. Currently the program supports downloading tweets based on geographical locations. The user needs to provide the NE and SW latitude and longitude coordinates which serve to define a bounding box from which the tweets are downloaded. Based on specific requirements, the program can be upgraded to download tweets by keywords, hashtags, usernames, et cetera. In other words, tweets can be downloaded according to any of the filters offered by the Twitter API. Research has shown that the Twitter API, while not a random sample of the full corpus of Twitter posts, can be used reliably for many types of topical analyses, especially those based on bounded geography (Morstatter et al. 2013).

Based on current requirements, the version of the software used in this book downloads the following fields of a tweet:

© The Editor(s) (if applicable) and The Author(s) 2016

J.B. Hollander et al., *Urban Social Listening*,
DOI 10.1057/978-1-137-59491-4

User ID, Username, Text, Longitude, Latitude, Language, Created at.

The program can be tweaked to download a great deal more information about each tweet.

## Analyzing for Sentiments

The sentiment analyzer currently scans each tweet for keywords defined in a dictionary, rated according to their sentiment with an integer. The program comes with a default dictionary based on AFINN.

The AFINN dictionary was developed by Finn Årup Nielsen, and ranks words on an ordinal scale ranging from +5 to –5. For example, "abusive" is given a score of –3, while "satisfied" is given a score of +2. The latest version of AFINN has 2477 words, and is capable of capturing variants of words such as recognizing "loooooove" as "love." It has been used in multiple research studies to date, including an analysis of tweets emanating from New Bedford, MA between February 9, 2014 and April 3, 2014 (Hollander et al. 2014), identification of anti-vaccine sentiments from tweets (Brooks 2014), evaluation of more than 5000 advertisements in business magazines (Abrahams et al. 2013), and as part of a model predicting fluctuations in global currency markets (Jin et al. 2013).

The score of each sentiment word is summed up for every tweet and the net score gives a measure of the sentiment present in a dataset. The analysis can be performed in conjunction with parameters that allow tweets to be filtered by date-time stamp, presence of keywords, and other factors. This is a useful feature to have, especially if you wish to analyze tweets by topic/hashtags or other indicators. In addition, the program also allows any text file to be analyzed with an inbuilt text analyzer, which is similar to the tweet analyzer.

The sentiment analyzer allows for scanning by wildcards, whereby defining words with a '*' following a sequence of characters and a corresponding score enables the program to score all words with that pattern to be scored the same. For example, kind* scans for kind, kindly, kinder, etc. and assigns the same score to every iteration of the associated sentiment.

## Future Additions

The program can be upgraded to mine data from all social networking and rating sites, which provide public APIs to access their data, and can be customized to analyze them. Possible additions include support for Yelp, Foursquare, and Facebook.

# Appendix 2

T-tests between proportions of positive tweets in Portuguese results

| | *Eastern MA* | *Brockton* | *Fall River* | *New Bedford* | *Lisbon* | *Sao Paulo* | *Harvard & MIT* | *Lowell* |
|---|---|---|---|---|---|---|---|---|
| Eastern MA | | | | | | | | |
| Brockton | Z-Score -0.1427, p-value 0.44433 | | | | | | | |
| Fall River | Z-Score 0.0336, p-value 0.48803 | Z-Score -0.0775, p-value 0.46812 | | | | | | |
| New Bedford | Z-Score 0.0002, p-value 0.5 | Z-Score -0.047, p-value 0.48006 | Z-Score -0.0249, p-value 0.49202 | | | | | |
| Lisbon | Z-Score -4.5619, p-value 0 | Z-Score 4.3704, p-value is 0 | Z-Score 1.2699, p-value 0.10204 | Z-Score 1.3175, p-value 0.09342 | | | | |
| Sao Paulo | Z-Score -7.8284, p-value 0 | Z-Score 7.558, p-value 0 | Z-Score -1.797, p-value 0.03593 | Z-Score -1.8819, p-value 0.03005 | Z-Score -3.0997, p-value 0.00097 | | | |

(*continued*)

**Table 1.4** (continued)

| | *Eastern MA* | *Brockton* | *Fall River* | *New Bedford* | *Lisbon* | *Sao Paulo* | *Harvard & MIT* | *Lowell* |
|---|---|---|---|---|---|---|---|---|
| Harvard & MIT | Z-Score 1.4539, p-value 0.07353 | Z-Score -1.4883, p-value 0.06811 | Z-Score -1.1421, p-value 0.12714 | Z-Score -1.1911, p-value 0.11702 | Z-Score 2.4416, p-value 0.00734 | Z-Score -2.866, p-value 0.00205 | | |
| Lowell | Z-Score -0.0109, p-value 0.49601 | Z-Score -0.0447, p-value 0.48405 | Z-Score -0.0249, p-value is 0.49202 | Z-Score 0.0075, p-value 0.49601 | Z-Score 1.562, p-value 0.05938 | Z-Score -2.2493, p-value 0.01222 | Z-Score -1.2664, p-value 0.10204 | |

# Appendix 3

## T-tests between proportions of positive tweets in English results

| | *Eastern MA* | *Brockton* | *Fall River* | *New Bedford* | *Lisbon* | *Sao Paulo* | *Harvard & MIT* | *Lowell* |
|---|---|---|---|---|---|---|---|---|
| Eastern MA | | | | | | | | |
| Brockton | Z-Score 0.3874, p-value 0.34827 | | | | | | | |
| Fall River | Z-Score 4.6413, p-value 0 | Z-Score 4.4253, p-value 0 | | | | | | |
| New Bedford | Z-Score 8.0452, p-value 0 | Z-Score 7.7692, p-value 0 | Z-Score 2.0976, p-value 0.01786 | | | | | |
| Lisbon | Z-Score 12.2877, p-value 0 | Z-Score 12.2002, p-value 0 | Z-Score 9.8446, p-value 0 | Z-Score 8.9547, p-value 0 | | | | |
| Sao Paulo | Z-Score 34.6994, p-value 0 | Z-Score 33.656, p-value 0 | Z-Score 15.3997, p-value 0 | Z-Score 13.9761, p-value 0 | Z-Score -3.8138, p-value 7E-05 | | | |
| Harvard & MIT | Z-Score -9.0885, p-value 0 | Z-Score -9.1635, p-value 0 | Z-Score -10.4421, p-value 0 | Z-Score -12.04, p-value 0 | Z-Score -15.4439, p-value 0 | Z-Score -20.5418, p-value 0 | | |
| Lowell | Z-Score 5.9444, p-value 0 | Z-Score 5.6902, p-value 0 | Z-Score 0.6054, p-value 0.27093 | Z-Score -1.5762, p-value 0.05705 | Z-Score -9.6709, p-value 0 | Z-Score -15.8637, p-value 0 | Z-Score 11.0646, p-value 0 | |

# Appendix 4

| *Demographic variable* | *Average for Brazilian and Portuguese ancestry* | *Average for four Gateway Cities* |
|---|---|---|
| % with high school diploma or higher | 70.4% | 75.0% |
| Median household income | 46,155 | 41,922 |
| % under 18 | 24.1% | 23.7% |
| % 65 and over | 7.0% | 13.0% |
| % Foreign born | 53.8% | 22.2% |
| Population | 23,908 | 96,681 |

*Sources:* United States Census Bureau (2015f, g, h, i, j)

# Appendix 5

## *2013 Tweet Data*

| *City* | *Positive score* | *Negative score* | *Overall score* | *Average score per tweet* | *Sentiment containing tweets* | *Total tweets* |
|---|---|---|---|---|---|---|
| Atlanta | 5,306,108 | -4,334,415 | 971,693 | 0.211 | 2,526,884 | 4,608,671 |
| Houston | 6,383,987 | -5,215,843 | 1,168,144 | 0.200 | 3,101,918 | 5,827,597 |
| Indianapolis | 2,221,306 | -1,654,203 | 567,103 | 0.314 | 1,035,167 | 1,804,805 |
| New Orleans | 2,603,198 | -2,359,788 | 243,410 | 0.098 | 1,336,069 | 2,492,313 |
| Newark | 3,626,840 | -2,946,019 | 680,821 | 0.206 | 1,735,069 | 3,305,276 |
| Providence | 640,322 | -534,970 | 105,352 | 0.189 | 309,133 | 556,671 |
| St. Louis | 1,744,355 | -1,425,904 | 318,451 | 0.213 | 842,460 | 1,494,820 |
| Washington | 2,950,291 | -2,514,422 | 435,869 | 0.160 | 1,462,535 | 2,731,275 |

| *City* | *Positive sentiment tweets* | *Negative sentiment tweets* | *% Tweets w/ sentiment* | *% Positive sentiment tweets* | *% Negative sentiment tweets* | *% Difference* |
|---|---|---|---|---|---|---|
| Atlanta | 1,735,626 | 1,274,876 | 54.83 | 37.66 | 27.66 | 10.00 |
| Houston | 2,095,748 | 1,591,752 | 53.23 | 35.96 | 27.31 | 8.65 |
| Indianapolis | 723,960 | 514,026 | 57.36 | 40.11 | 28.48 | 11.63 |
| New Orleans | 879,760 | 704,040 | 53.61 | 35.30 | 28.25 | 7.05 |
| Newark | 1,176,198 | 885,125 | 52.49 | 35.59 | 26.78 | 8.81 |
| Providence | 208,508 | 160,536 | 55.53 | 37.46 | 28.84 | 8.62 |
| St. Louis | 576,301 | 429,502 | 56.36 | 38.55 | 28.73 | 9.82 |
| Washington | 983,297 | 751,414 | 53.55 | 36.00 | 27.51 | 8.49 |

## *2015 Tweet Data*

| *City* | *Positive score* | *Negative score* | *Overall score* | *Average score per tweet* | *Sentiment containing tweets* | *Total tweets* |
|---|---|---|---|---|---|---|
| Atlanta | 157,366 | -138,664 | 18,702 | 0.126 | 76,819 | 149,010 |
| Houston | 612,865 | -584,293 | 28,572 | 0.047 | 313,556 | 609,335 |
| Indianapolis | 133,556 | -106,406 | 27,150 | 0.235 | 63,268 | 115,311 |
| Manchester | 18,840 | -12,236 | 6604 | 0.433 | 8311 | 15,259 |
| New Britain | 8654 | -7673 | 981 | 0.131 | 4166 | 7487 |
| New Orleans | 156,535 | -170,049 | -13,514 | -0.074 | 89,875 | 183,195 |
| Newark | 61,741 | -57,615 | 4126 | 0.069 | 31,007 | 60,035 |
| Providence | 34,492 | -28,477 | 6015 | 0.181 | 17,132 | 33,288 |
| St. Louis | 43,255 | -34,343 | 8912 | 0.228 | 20,356 | 39,078 |
| Stamford | 10,796 | -7133 | 3663 | 0.376 | 5017 | 9741 |
| Washington | 239,492 | -210,311 | 29,181 | 0.123 | 120,351 | 237,626 |
| Waterbury | 17,660 | -20,622 | -2962 | -0.177 | 9504 | 16,769 |

| *City* | *Positive sentiment tweets* | *Negative sentiment tweets* | *% Tweets w/ sentiment* | *% Positive sentiment tweets* | *% Negative sentiment tweets* | *% Difference* |
|---|---|---|---|---|---|---|
| Atlanta | 52,441 | 38,811 | 51.55 | 35.19 | 26.05 | 9.15 |
| Houston | 204,731 | 168,041 | 51.46 | 33.60 | 27.58 | 6.02 |
| Indianapolis | 43,360 | 31,883 | 54.87 | 37.60 | 27.65 | 9.95 |
| Manchester | 6025 | 3966 | 54.47 | 39.48 | 25.99 | 13.49 |
| New Britain | 2821 | 2199 | 55.64 | 37.68 | 29.37 | 8.31 |
| New Orleans | 57,260 | 48,903 | 49.06 | 31.26 | 26.69 | 4.56 |
| Newark | 20,409 | 16,758 | 51.65 | 34.00 | 27.91 | 6.08 |
| Providence | 11,554 | 8873 | 51.47 | 34.71 | 26.66 | 8.05 |
| St. Louis | 14,053 | 10,227 | 52.09 | 35.96 | 26.17 | 9.79 |
| Stamford | 3547 | 2310 | 51.50 | 36.41 | 23.71 | 12.70 |
| Washington | 80,763 | 61,817 | 50.65 | 33.99 | 26.01 | 7.97 |
| Waterbury | 6033 | 5481 | 56.68 | 35.98 | 32.69 | 3.29 |

| *City* | *Median income* | *% Under 18* | *College graduation %* | *Pop. change 1970–2010* | *Foreign born %* | *Mean travel time to work* | *Median value of owned homes* | *% Persons below poverty* |
|---|---|---|---|---|---|---|---|---|
| Atlanta | $46,631 | 19.4 | 46.8 | -76,970 | 7.7 | 25.1 | $210,000 | 25.0 |
| Houston | $45,010 | 25.9 | 29.2 | 867,461 | 28.3 | 25.9 | $123,900 | 22.9 |
| Indianapolis | $41,962 | 25.0 | 27.3 | 75,821 | 8.7 | 22.6 | $118,000 | 20.9 |
| New Orleans | $37,146 | 21.3 | 33.7 | -249,642 | 5.9 | 23.0 | $183,700 | 27.3 |
| Newark | $33,960 | 25.6 | 12.7 | -104,790 | 27.2 | 32.3 | $243,200 | 29.1 |
| Providence | $37,632 | 23.4 | 28.5 | -1171 | 30.0 | 21.4 | $196,300 | 29.0 |
| St. Louis | $34,582 | 21.2 | 29.6 | -302,942 | 6.7 | 23.9 | $119,200 | 27.4 |
| Washington | $65,830 | 17.2 | 52.4 | -154,787 | 13.8 | 29.7 | $445,200 | 18.6 |

*Source:* US Census Quick Facts

Atlanta: http://quickfacts.census.gov/qfd/states/13/1304000.html

Houston: http://quickfacts.census.gov/qfd/states/48/4835000.html

Indianapolis: http://quickfacts.census.gov/qfd/states/18/1836003.html

New Orleans: http://quickfacts.census.gov/qfd/states/22/2255000.html

Newark: http://quickfacts.census.gov/qfd/states/34/3451000.html

Providence: http://quickfacts.census.gov/qfd/states/44/4459000.html

St. Louis: http://quickfacts.census.gov/qfd/states/29/2965000.html

Washington: http://quickfacts.census.gov/qfd/states/11000.html

# Bibliography

Batty, M. (2012). Smart cities, big data. *Environment and Planning B, 39*, 191–193.

Batty, M. (2013). Big data, smart cities, and city planning. *Dialogues in Human Geography, 3*(3), 274–279.

Bifet, A., & Frank, E. (2010). Sentiment knowledge discovery in twitter streaming data. In *Discovery Science*. Berlin/Heidelberg: Springer.

Delen, D. (2009). Predicting movie Box-office receipts using SPSS Clementine data mining software. In R. Nisbet, J. Elder, & G. Miner (Eds.), *Handbook of statistical analysis and data mining applications* (pp. 391–415). Amsterdam: Elsevier. ISBN 978-0-12-374765-5.

Fung, A. (2006). Varieties of participation in complex governance. *Public Administration Review, 66*(1), 66–75.

Hempel, L. C. (1999). Conceptual and analytical challenges in building sustainable communities. In D. A. Mazmanian & M. E. Kraft (Eds.), *Towards sustainable communities: Transition and transformations in environmental policy* (pp. 33–62). Cambridge, MA: The MIT Press.

Hollander, J., Graves, E., & Levanthal, T. (2014). Using big data to study urban sentiments: Twitter data vs. Published Meeting Minutes. Tufts University.

Hollander, J. B., & Németh, J. (2011). The bounds of smart decline: A foundational theory for planning shrinking cities. *Housing Policy Debate, 21*(3), 349–367.

Hughes, J. (2014, November 17). Can Government mine tweets to assess public opinion?. Government Technology.

Ivanov, I. (2013). *Advanced social media analysis.* Dissertation. EPFL.

Johnson, M. P. (Ed.). (2011). *Community-based operations research: Decision modeling for local impact and diverse populations.* New York: Springer.

© The Editor(s) (if applicable) and The Author(s) 2016

J.B. Hollander et al., *Urban Social Listening*,
DOI 10.1057/978-1-137-59491-4

Keeney, R. L. (1992). *Value-focused thinking: A path to creative decisionmaking.* Cambridge, MA: Harvard University Press.

Kouloumpis, E., Wilson, T., & Moore, J. (2011). Twitter sentiment analysis: The good the bad and the omg! *ICWSM, 11*, 538–541.

Larson, R., & Odoni, A. (2007). *Urban operations research* (2nd ed.). Charlestown: Dynamic Ideas, Inc.

Leetaru, K. (2012). *Data mining methods for the content analyst: An introduction to the computational analysis of content.* New York/London: Routledge.

Leland, J. (2007, December 25). Officials say they are falling behind on mortgage fraud cases. *The New York Times.*

Leventhal, T., & Brooks-Gunn, J. (2011). Changes in neighborhood poverty from 1990 to 2000 and youth's problem behaviors. *Developmental Psychology, 47*(6), 1680–1698.

Logan, J. R., & Molotch, H. L. (1987). *Urban fortunes: The political economy of place.* Berkeley: University of California Press.

Lucy, W. H., & Phillips, D. L. (2000). *Confronting suburban decline: Strategic planning for metropolitan renewal.* Washington, DC: Island Press.

Malczewski, J. (1999). *GIS and multicriteria analysis.* New York: Wiley.

Midgley, G., & Ochoa-Arias, A. (Eds.). (2004). *Community operational research: OR and systems thinking for community development.* New York: Academic/Plenum.

Morstatter, F., Pfeffer, J., & Liu, H. (2013). When is it biased? Assessing the representativeness of Twitter's streaming API. *International World Wide Web Conference Committee (IW3C2)*, 555–556.

Nielsen, F. Å. (2011, March). AFINN. Informatics and Mathematical Modelling. Technical University of Denmark.

Offenhuber, D., Lee, D., Wolf, M. I., Phithakkitnukoon, S., Biderman, A., & Ratti, C. (2012). Putting matter in place: Measuring tradeoffs in waste disposal and recycling. *Journal of the American Planning Association, 78*(2), 173–196.

Pachamanova, D. A., & Fabozzi, F. J. (2014). Recent trends in equity portfolio construction analytics. *The Journal of Portfolio Management, 40*(3), 137–151.

Paek, H-J., Hove, T., & Cole, R. T. (2013). A multimethod approach to evaluating social media campaign effectiveness. *System Sciences (HICSS), 2013 46th Hawaii International Conference,* IEEE.

Pak, A., & Paroubek, P. (2010). Twitter as a Corpus for sentiment analysis and opinion mining. *LREC.*

Rosenhead, J., & Mingers, J. (2001). *Rational analysis for a problematic world, revisited: Problem structuring methods for complexity, uncertainty and conflict* (2nd ed.). Chichester: Wiley.

Salim, F. D. (2012). Probing streets and the built environment with ambient and community sensing. *Journal of Urban Technology, 19*(2), 47–67.

SentiStrength (2015). About SentiStrength. http://sentistrength.wlv.ac.uk/#About. Accessed 31 Mar 2015.

Stokey, E., & Zeckhauser, R. (1978). *A primer for policy analysis.* New York: W.W. Norton & Company.

Timmermans, H. (1997). *Decision support systems in urban planning.* London: E&F SPON.

Tuhus-Dubrow, R. (2014, October 21). Will Twitter revolutionize how cities plan for the future?. NextCity.org

UMassDartmouth (2008). The Gateway Cities compact for community and economic development. http://www.umassd.edu/media/umassdartmouth/seppce/centerforpolicyanalysis/urbaninitiative/gatewaycities/signedcompact.pdf. Accessed 6 Nov 2014.

United States Census Bureau (2015f). DP02 selected social characteristics in the United States. 2006–2010 American Community Survey Selected Population Tables.

United States Census Bureau (2015g). B01001 Sex by age. 2006–2010 American Community Survey Selected Population Tables.

United States Census Bureau (2015h). B19013 Median household income in the past 12 months (in 2010 inflation-adjusted dollars). 2006–2010 American Community Survey Selected Population Tables.

United States Census Bureau (2015i). CP02 comparative social characteristic in the United States. 2013 American Community Survey 1-year Estimates.

United States Census Bureau (2015j). B05002 Place of birth by citizenship status. 2006–2010 American Community Survey Selected Population Tables.

Winston, W. L., & Albright, S. C. (2012). *Practical management science* (4th ed.). Mason: South-Western/CENGAGE Learning.

# Index



© The Editor(s) (if applicable) and The Author(s) 2016
J.B. Hollander et al., *Urban Social Listening*,
DOI 10.1057/978-1-137-59491-4